# UAV Cooperative Decision and Control

## Advances in Design and Control

SIAM's Advances in Design and Control series consists of texts and monographs dealing with all areas of design and control and their applications. Topics of interest include shape optimization, multidisciplinary design, trajectory optimization, feedback, and optimal control. The series focuses on the mathematical and computational aspects of engineering design and control that are usable in a wide variety of scientific and engineering disciplines.



## Series Volumes

Shima, Tal and Rasmussen, Steven, eds., *UAV Cooperative Decision and Control: Challenges and Practical Approaches*
Speyer, Jason L. and Chung, Walter H., *Stochastic Processes, Estimation, and Control*
Krstic, Miroslav and Smyshlyaev, Andrey, *Boundary Control of PDEs: A Course on Backstepping Designs*
Ito, Kazufumi and Kunisch, Karl, *Lagrange Multiplier Approach to Variational Problems and Applications*
Xue, Dingyü, Chen, YangQuan, and Atherton, Derek P., *Linear Feedback Control: Analysis and Design with MATLAB*
Hanson, Floyd B., *Applied Stochastic Processes and Control for Jump-Diffusions: Modeling, Analysis, and Computation*
Michiels, Wim and Niculescu, Silviu-Iulian, *Stability and Stabilization of Time-Delay Systems: An Eigenvalue-Based Approach*
Ioannou, Petros and Fidan, Barış, *Adaptive Control Tutorial*
Bhaya, Amit and Kaszkurewicz, Eugenius, *Control Perspectives on Numerical Algorithms and Matrix Problems*
Robinett III, Rush D., Wilson, David G., Eisler, G. Richard, and Hurtado, John E., *Applied Dynamic Programming for Optimization of Dynamical Systems*
Huang, J., *Nonlinear Output Regulation: Theory and Applications*
Haslinger, J. and Mäkinen, R. A. E., *Introduction to Shape Optimization: Theory, Approximation, and Computation*
Antoulas, Athanasios C., *Approximation of Large-Scale Dynamical Systems*
Gunzburger, Max D., *Perspectives in Flow Control and Optimization*
Delfour, M. C. and Zolésio, J.-P., *Shapes and Geometries: Analysis, Differential Calculus, and Optimization*
Betts, John T., *Practical Methods for Optimal Control Using Nonlinear Programming*
El Ghaoui, Laurent and Niculescu, Silviu-Iulian, eds., *Advances in Linear Matrix Inequality Methods in Control*
Helton, J. William and James, Matthew R., *Extending $H^\infty$ Control to Nonlinear Systems: Control of Nonlinear Systems to Achieve Performance Objectives*

# UAV Cooperative Decision and Control

## Challenges and Practical Approaches


Edited by

**Tal Shima**

Department of Aerospace Engineering
Technion–Israel Institute of Technology, Haifa

**Steven Rasmussen**

Control Science Center of Excellence
Miami Valley Aerospace LLC
Wright-Patterson Air Force Base, Ohio



siam.

Society for Industrial and Applied Mathematics
Philadelphia

10 9 8 7 6 5 4 3 2 1



Figure 1.1 appears courtesy of the United State Air Force's Aeronautical System Center History Office.
Figure 1.2 appears courtesy of the National Museum of the United States Air Force.
Figure 1.3 appears courtesy of the United States Air Force archives.
Figures 1.5 and 1.6 appear courtesy of NASA.
Figure 3.1 appears with permission of Lockheed Martin.

**Library of Congress Cataloging-in-Publication Data**

UAV cooperative decision and control : challenges and practical approaches / edited by Tal Shima, Steven Rasmussen.
p. cm. – (Advances in design and control ; no. 18)
Includes index.
ISBN 978-0-898716-64-1
1. Drone aircraft–Control systems. 2. Drone aircraft–Automatic control. 3. Task analysis. 4. Decision making. I. Shima, Tal. II. Rasmussen, Steven J.
UG1242.D7U27 2008
358.4'5–dc22

2008032459

# Contributors

Nicola Ceccarelli
*National Research Council*
*Control Science Center of Excellence*
*Air Force Research Laboratory*
*Wright-Patterson AFB, OH* 45433, *USA*

Phillip Chandler
*Control Science Center of Excellence*
*Air Force Research Laboratory*
*Wright-Patterson AFB, OH* 45433, *USA*

David Jacques
*Department of Aeronautics and Astronautics*
*U.S. Air Force Institute of Technology*
*Wright-Patterson AFB, OH* 45433, *USA*

Brian Kish
*U.S. Air Force Institute of Technology*
*Wright-Patterson AFB, OH* 45433, *USA*

Meir Pachter
*Department of Electrical and Computer Engineering*
*U.S. Air Force Institute of Technology*
*Wright-Patterson AFB, OH* 45433, *USA*

Steven Rasmussen
*Miami Valley Aerospace LLC*
*Control Science Center of Excellence*
*Air Force Research Laboratory*
*Wright-Patterson AFB, OH* 45433, *USA*

Corey Schumacher
*Control Science Center of Excellence*
*Air Force Research Laboratory*
*Wright-Patterson AFB, OH* 45433, *USA*

Tal Shima
*Department of Aerospace Engineering*
*Technion—Israel Institute of Technology*
32000, *Haifa, Israel*

# Contents

# List of Figures

# List of Tables

# Foreword

The research that forms the foundation for this book was accomplished at the Air Force Research Laboratory's (AFRL) Control Science Center of Excellence (CSCOE). As the director of the CSCOE, I'm pleased to introduce this body of work. One of the tasks of the CSCOE is to perform research into controlling multiple unmanned aerial vehicles (UAVs). We started this task in the late 1990s with research into *cooperative rendezvous*, which was followed with topics including *cooperative wide area search and destroy*, *cooperative intelligence surveillance and reconnaissance*, *cooperative electronic attack*, and *cooperative operations in urban terrain*. Throughout this work our researchers strived to extract the essential attributes of each of the control problems so these attributes can be applied on future UAV systems. Over the years we have gained a lot of insight into the problem of controlling multiple UAVs and that is the *raison d'être* for this book.

Wright-Patterson Air Force Base has been the center of aerospace technology development since the Wright Brothers first developed airplanes at their nearby Dayton, Ohio, bicycle shop and test flew their designs at Huffman Prairie, now a part of the base. The AFRL, headquartered at Wright-Patterson, maintains this tradition as one of the largest complexes in the world dedicated to excellence in the aerospace sciences. The CSCOE, a part of AFRL's Air Vehicles Directorate, is the laboratory's leader in the development of the control technologies necessary to maintain the United States Air Force as the preeminent aerospace power. The CSCOE is staffed by a highly professional cadre of award-winning civil service scientists and Air Force officers who form the technical core of the center's competencies. This core is augmented by numerous visiting scientists who provide a fresh perspective to the center's tasks. This partnership ensures that the center maintains its technical preeminence. The center's research tasks cover a wide spectrum of aerospace science applications. From air vehicles to transatmospheric vehicles, including both manned and unmanned aerial vehicles, the CSCOE is tasked with developing and transitioning advanced control technologies for all aspects of the 21st-century air and space force.

In the future, UAVs will operate in teams to autonomously perform complex, cooperative tasks such as destruction of enemy threats and time-critical targets. As such, these vehicles will require algorithms for making decisions that meet team goals and mission objectives. Cooperative control of UAVs is a complex problem that is dominated by task coupling, limited information, and a high degree of uncertainty. We are developing algorithms for real-time multiple-task assignment with complex task constraints. Additionally, we are developing decentralized decision and control algorithms to provide robustness and flexibility. As these teams of UAVs will operate in uncertain and adversarial environments, communications will be critical. Delays can create instability, limit cycles, and cause loss of cohesion as a team. We are developing metrics and control strategies that enable us to maintain or gracefully degrade coordinated team performance despite arbitrary communications delays and highly dynamic events. In addition, information theory and game theory are being pursued to address uncertainty about the environment and the adversary in battlespace operations with heterogeneous vehicles. Interfacing with the operator and integration into a battle network are important areas of continuing research. An exciting new field of investigation is the cooperative control of a team of small and micro air vehicles in urban environment missions. This scenario presents significant challenges with the combination of high uncertainty, with limited sensors, processing, and communication. While we have made many advances, we have barely scratched the surface in intelligent cooperative control and its potential for the future.

I would like to recognize the contributions of other directorates, specifically the Munitions Directorate and Human Effectiveness Directorate. During this period several long-term fellows, professors on sabbatical, on-site contractors, and our collaborative center have made many contributions to the research. Many summer faculty, summer students, and visitors have added a broad range of expertise and insight. And of course I would like to thank the Air Vehicles Directorate and Air Force Office of Scientific Research, whose support made this work possible.

Siva Banda, Director
Control Science Center of Excellence
Air Force Research Laboratory
Wright-Patterson Air Force Base, Ohio

# Preface

The use of unmanned aerial vehicles (UAVs) for various military missions has received growing attention in the last decade. Apart from the obvious advantage of not placing human life at risk, the lack of a human pilot enables significant weight savings and lower costs. UAVs also provide an opportunity for new operational paradigms. To realize these advantages, UAVs must have a high level of autonomy and preferably work cooperatively in groups. Exchanging information within these groups can greatly improve their capability.

In this context, a concentrated research effort has been conducted in recent years to develop novel cooperative decision and control algorithms. These algorithms deal with the problem of commanding multiple UAVs to cooperatively perform multiple tasks. The need is to assign specific tasks and flyable trajectories to each vehicle to maximize the group performance. The problem is challenging because the assignment task involves binary decision variables, while the path optimization involves continuous ones, and both are coupled. Also, to allow implementation, the developed algorithms must be solved in real time, possibly under uncertainty and communication constraints, and must be robust to failures.

This book provides an authoritative reference on U.S. Air Force relevant UAV cooperative decision and control problems and the means available to solve them. The book is aimed at helping practitioners, academicians, and students alike to better understand what cooperative decision and control is and its applications and methods for implementing algorithms that make cooperative UAV operations possible. The approach of this book is to present the UAV cooperative decision and control problem in a manner that abstracts the challenges from the concrete problems, making it possible to leverage the solution methods over a broader range of applications. To help researchers new to the field, and those already in the field, a thorough description of the problem and its challenges is presented. Solution algorithms that have recently been developed using various approaches will be presented, providing a baseline for further study.

To further enhance the subject matter, a multiple UAV simulation test bed, MultiUAV2, accompanies the text, making it possible for researchers to investigate new cooperative control strategies. MultiUAV2 is a Simulink–MATLAB–C++-based simulation that is capable of simulating multiple unmanned aerial vehicles that cooperate to accomplish predefined missions.

The outline of the book is as follows.

Chapter 1 gives an introduction to the cooperative decision and control problem. The road to autonomous control of UAVs is described and classes of cooperation are defined. Representative scenarios are used to describe and define the problem and its challenges.

Chapter 2 provides an in-depth study of the challenges associated with cooperative control of multiple UAVs. A taxonomy of teams is provided and the complexity of cooperative operation is analyzed. An overview of algorithms that could be used for cooperative control of UAV teams is presented.

Chapter 3 describes the baseline scenario of multiple UAVs performing multiple tasks on multiple targets. A linear programming formulation known as a capacitated transshipment assignment problem is then provided. This method is used to assign at each stage a single task to a single UAV. An iterative application of this algorithm, allowing the prosecution of multiple tasks on multiple targets, is then presented.

Chapter 4 presents three methods to assign multiple tasks to multiple UAVs in one step. These methods are mixed integer linear programming (MILP), tree search, and genetic algorithms (GA). Each has its respective pros and cons, which are described. Simulation results are also provided.

Chapter 5 studies a further complication of the cooperative problem, in which the multiple tasks on each target have to be performed simultaneously. This results in strict timing constraints for each task that must be addressed. The application of MILP and GA to this problem is studied and compared.

Chapter 6 deals with the cooperation of UAVs under communication delays. These delays may produce different information sets for the different UAVs in the group, leading to uncoordinated assignments. A decision-estimation architecture enhancing cooperation in such a scenario is presented. For the estimation process, communication and computation efficient algorithms are provided.

Chapter 7 presents effectiveness measures derived for UAV operations in uncertain environments. These measures provide a foundation for the rational evaluation of cooperation rules of engagement and for the effectiveness of aerial munitions, tactical UAVs, and general sensor craft.

Appendix A describes the MultiUAV2 simulation and its operation.

Appendix B details the UAV path planning problem and Dubins' optimal trajectories.

All the material and experience contained in the book comes from research performed at the U.S. Air Force Research Laboratory's Control Science Center of Excellence within the Air Vehicles Directorate located at Wright-Patterson Air Force Base. The book is the result of almost a decade of research performed by in-house staff and contractors, collaborators from the Air Force Institute of Technology located at Wright-Patterson AFB, and visiting scientists.

As editors of this book, we would like to pay special gratitude to all the authors who contributed much time and effort to this endeavor. Special thanks go to Dr. Siva Banda, head of the Control Science Center of Excellence, for his vision and support of the research presented in this book. We also wish to extend our appreciation for the comments and suggestions of the reviewers, especially Prof. Anouck Girard, University of Michigan at Ann Arbor, who performed an extensive review on an early version of this manuscript. Last, we wish to thank Ms. Elizabeth Greenspan, the acquisitions editor of the Society for Industrial and Applied Mathematics, for her support and patience throughout the entire publication process.

January 2008

TAL SHIMA, Haifa, Israel
STEVEN RASMUSSEN, Wright-Patterson Air Force Base, Ohio, USA

# Acronyms

| | |
|---|---|
| 6DOF | Six-Degree-of-Freedom |
| ACL | Autonomous Control Level |
| ATR | Automatic Target Recognition |
| BFS | Best-First Search |
| CMTE | Cooperative Moving Target Engagement |
| CS | Computer Science |
| CTAP | Capacitated Transhipment Assignment Problem |
| DFS | Depth-First Search |
| DM | Decision Maker |
| ECAV | Electronic Combat Aerial Vehicle |
| EFS | Embedded Flight Software |
| GA | Genetic Algorithm |
| GMTI | Ground Moving Target Indication |
| GNC | Guidance Navigation and Control |
| GPS | Global Positioning System |
| IF | Information Filter |
| ISR | Intelligence, Surveillance, and Reconnaissance |
| KF | Kalman Filter |
| LADAR | Laser Detection and Ranging |
| LOS | Line of Sight |
| MAV | Micro Aerial Vehicle |

| | |
|---|---|
| MILP | Mixed Integer Linear Programming |
| OR | Operations Research |
| POMDP | Partially Observable Markov Decision Process |
| RCI | Redundant Centralized Implementation |
| ROC | Receiver Operating Characteristic |
| SEAD | Suppression of Enemy Air Defenses |
| UAV | Unmanned Aerial Vehicle |
| VCVS | Variable Configuration Vehicle Simulation |
| VRP | Vehicle Routing Problem |
| WASM | Wide-Area Search Munition |

# Chapter 1

# Introduction

***Steven Rasmussen, Tal Shima, and Phillip Chandler***

*Great discoveries and improvements invariably involve the cooperation of many minds.*
—Alexander Graham Bell

The use of unmanned vehicles for various military and civilian missions in the air, in the sea, in space, and on the ground has received growing attention in the last decade. Apart from the obvious advantage of not placing human life in harm's way, the lack of an on-board human operator enables longer endurance. Moreover, working in groups presents the opportunity for new operational paradigms. In this book, we will concentrate on the cooperative decision and control problem associated with the operation of a group of unmanned aerial vehicles (UAVs) against multiple ground targets. The concepts that will be presented are general and thus can be extended to other problems involving cooperative unmanned systems.

## 1.1 Road to Cooperative Control

### 1.1.1 Some History on Unmanned Systems

In building unmanned systems, the problem has always been one of technological capability. To construct an effective unmanned system, one must implement means of controlling the vehicle in an uncertain environment throughout the mission. Controlling the vehicle includes navigation as well as weapon employment. One of the earliest ideas, developed by the Austrians and employed during their assault on Venice in 1849, was to use an unmanned balloon to float over enemy territory and drop bombs, which exploded on impact [1]. One obvious operational problem was the difficulty in controlling the trajectory of the balloons.

**Figure 1.1:** Kettering Aerial Torpedo. *(Courtesy of Aeronautical System Center History Office.)*

The operational range was also severely limited since for weapon employment the bombs were released through the use of an electrical charge over a copper wire.

At the beginning of World War I, the Kettering (Bug) Aerial Torpedo (Fig. 1.1) was developed [12]. It was designed to be launched from a rail, fly over enemy territory, and dive into enemy lines. It carried 180 pounds of high explosives which detonated when the vehicle hit the ground. Trajectory control of the Kettering Bug was open loop, based on the direction of launch and dependent on the winds en route. Automatic weapons employment was based on estimating the travel distance by counting the rotations of the propeller and then stopping the engine and jettisoning the wings at the calculated point.

During World War II, UAVs were used to attack both naval and land targets. They were also used for aerial reconnaissance and target practice. Manned aircraft were used to control the unmanned weapons through a radio link. Germany used weapons such as the Henschel Hs 293 Air-to-Ship Wireless Guided Bomb to attack a number of ships [8]. During the same time frame, the United States developed many different unmanned systems. One

**Figure 1.2:** Lockheed D-21B unmanned reconnaissance aircraft. *(Courtesy of the National Museum of the USAF.)*

noteworthy system was the Interstate TDR-1 Assault Drone [12]. This drone was radio controlled by a manned aircraft aided by a television camera carried on board the drone. This made it possible to precisely home in on the target during the terminal phase of the attack.

During the cold war, reconnaissance had a high priority and therefore many unmanned systems, such as the Lockheed D-21B (see Fig. 1.2) were developed to fly high-altitude strategic reconnaissance missions [7]. The Lockheed D-21B, developed in the early 1960s, was guided by an inertial navigation system to fly on a preprogrammed flight profile. In 1986, the United States procured its first tactical UAV, the Israeli-developed IAI/AAI RQ-2 Pioneer [4]. The Pioneer is still used for short-range reconnaissance (<100 nm) and is normally remotely piloted.

In the 1990s, new airframe construction techniques, the Global Positioning System (GPS), secure reliable communications, and other technological advancements made it possible to develop and deploy UAVs capable of performing worldwide reconnaissance missions with minimal manned intervention. Vehicles such as the General Atomics RQ-1 Predator and the Northrop Grumman RQ-4A Global Hawk (see Fig. 1.3) have sophisticated mission sensors that relay data to other assets in real time, but the vehicles themselves do not use the gathered information to autonomously make mission decisions. Instead, the ground operators make these decisions [4].

### 1.1.2 Autonomous Systems

Many of the early UAV systems, such as the Kettering Aerial Torpedo, operated open loop. Thus, any changes in the environment parameters (such as wind) could cause mission failure.

**Figure 1.3:** Northrop Grumman Global Hawk. *(Courtesy of the USAF.)*

In modern systems, navigation can be accomplished automatically, i.e., closed loop, but there are many other requirements that occur during mission execution, such as target selection and weapons employment, that necessitate system autonomy in making decisions. In the early 1980s, unmanned systems were designed with sufficient autonomy, within limits, to decide where to strike. For example, the Northrop AGM-136A Tacit Rainbow (see Fig. 1.4) was designed to be released from another aircraft, fly to a predefined area, loiter in that area until an enemy radar was energized, and then attack that radar. The Tacit Rainbow was designed with the autonomy to choose its target and the trajectory required to attack it.

From the late 1990s to the present, there has been significant interest in developing autonomous unmanned combat air vehicles. Note that the term *combat* refers to the vehicle's ability to use deadly force against the enemy. These vehicles, such as the Boeing X-45a (see Fig. 1.5), will have sophisticated sensors and weapons [4] and will be given the autonomy to make mission-related decisions. Such decisions can relate, for example, to the best route to a target or target queuing.

Depending on the application, there are many different ideas on how to measure the autonomy of a system. In [2], the autonomous control level (ACL) metric was introduced. This metric has 11 levels ranging from Remotely Piloted Vehicle to Fully Autonomous. To rank its autonomy, a system is rated in several areas, including Perception/Situational Awareness, Analysis/Coordination, Decision Making, and Capability. The ACLs are shown in Fig. 1.6, along with estimated rankings of the vehicles mentioned previously. Note that the Balloon Bomb and the Kettering Aerial Torpedo were given an ACL of 1. Both systems had more autonomy than the Henschel Hs 293 and the Interstate TDR-1, but neither of the earlier systems were as capable as the more modern systems.

**Figure 1.4:** Northrop AGM-136A Tacit Rainbow.

**Figure 1.5:** Boeing X-45a, unmanned combat air vehicle. *(Courtesy of NASA.)*

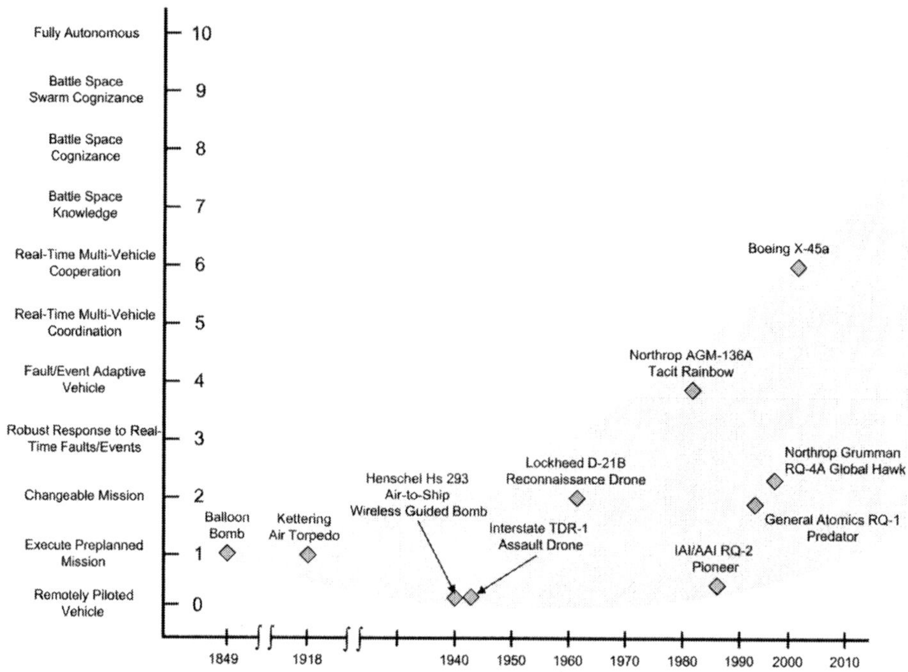


**Figure 1.6:** Estimated autonomous control levels.

# 1.2 Team Cooperation

## 1.2.1 Motivation

A loose collection of vehicles that have some objectives in common is a collaborative team. If the vehicles are working together to achieve a common objective, then they are a cooperative team. The main motivation for team cooperation stems from the possible synergy, as the group performance is expected to exceed the sum of the performance of the individual UAVs. Such cooperation, possible only if the UAVs have a high level of autonomy, should take advantage of the following capabilities available to the group.

### Global Information

Each UAV carries a payload enabling it to sense the environment. By sharing its sensor information with the rest of the group, via a communication network, the entire team can act based on this global situation awareness instead of the individually available local information. Such sensor cooperation is often termed network-centric.

### Resource Management

Each UAV can make decisions regarding its path and actions in the uncertain environment. Operating independently can cause certain targets to be overserviced by some of the UAVs,

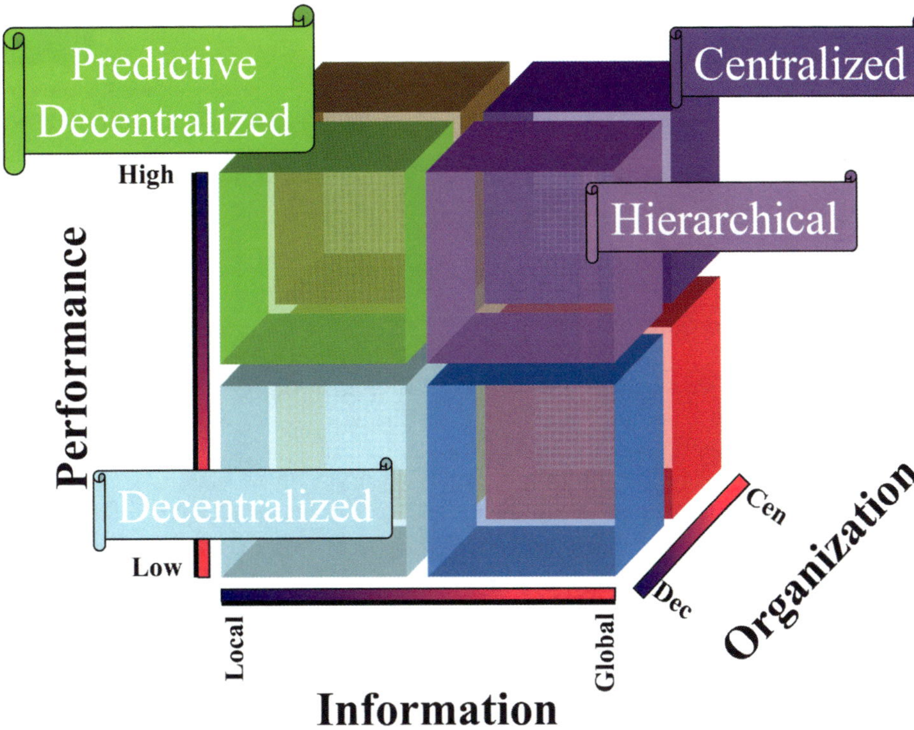


**Figure 1.7:** Team decision and control metrics.

while other targets may be underserviced. Having a cooperative decision algorithm allows efficient allocation of the overall group resources over the multiple targets.

#### Robustness

If the UAVs are independently dispersed over the terrain, then a failure of one of the vehicles may leave a subset of the target area uncovered. Through cooperation the team can reconfigure its distribution architecture to minimize the performance degradation to such expected failures.

### 1.2.2 Classes of Cooperative Controllers

Team cooperative decision and control controllers can be characterized in many ways, but a useful set of metrics is organization structure, information structure, and performance level. Controllers for individual problems can be dominated by one of these metrics but will contain elements of them all.

Fig. 1.7 shows different classes of cooperative decision and control controllers, with each axis representing one of the metrics mentioned above. Solutions for the lower part of the performance axis can be nominally thought of as preprogrammed or reactive, while the upper part can be thought of as more optimal control or adaptive control solutions. The right-hand part of the information axis has a high degree of global information and can be thought of as problems requiring measurement or sensor rich solutions, while the left-hand part consists primarily of local information and is therefore more dependent on predictive model solutions for good performance. The back part of the organizational axis is centralized and can be thought of as requiring command driven solutions, while the front part is more decentralized with an emphasis on cooperation solutions.

### Classical Centralized Controller

The classical centralized controller problem is situated in the upper-right back section of this cube. All the state information from the distributed vehicles or agents is sent to a centralized agent, where it is operated on by a large decision and control program. The resultant individual plans and assignments are disseminated to the respective vehicles to be executed. This approach can lack significant robustness, is computationally complex, and doesn't scale well. Scaling well means that controller complexity is roughly proportional to team size. The cooperative moving target engagement in Chapter 5 has coupled tasks that lead to a more centralized controller.

### Hierarchical Controller

The upper-right front section employs a decomposition, usually hierarchical. Now, only a subset of the vehicles send a subset of the local vehicle state to a team member. The team members have limited global information and send cost functions to a specific team agent so that individual assignments can be made that are beneficial to the team as a whole. While the vehicles have more autonomy and therefore robustness, it may be difficult to decompose the problem if there is appreciable task coupling. In general, this approach is more computable and scalable but is not as optimal as the centralized solution. The wide-area search munition problem in Chapter 3 is an example where the vehicles searching for targets are one subteam while the vehicles servicing targets that have already been found are another subteam.

### Decentralized Controller

The most decentralized problem, containing little if any global information, is situated in the lower-left front section of the cube. In the limit, there is no communication between the vehicles and information is only inferred about the other vehicles' objectives by measuring their actions through their sensors. In general, this approach leads to an average macro level of performance. For example, the team can maintain some loose cohesion and a nominal trajectory. The achievable performance here is low due to the absence of communication and little global coordinating information. This class of controller is not utilized in the remainder of the book because the solutions have tended more to centralized than decentralized control due to the difficulty of enforcing the dominant coupling constraints.

**Predictive Decentralized Controller**

The upper-left front section is simultaneously relatively decentralized and has good performance. Nominally this is a peer-to-peer team that may not have a team leader. The individual team members share a common objective. The vehicles are willing to accept an assignment that does not maximize their individual utility if it will maximize the overall team utility. There is a high dependence on state estimation and predictive models. The level of performance is dictated by the quality of the estimators. Arbitration is needed to resolve the conflicts if the tasks are coupled. The state prediction is addressed in Chapter 6. This class of controller was used on the electronic attack problem (see section 1.3.1) but is not presented in this book.

## 1.3 Cooperation Problems

The challenges in practical cooperative decision and control problems of the classes discussed above are driven by the requirements of proposed missions. Many different missions have been proposed that require cooperation between UAVs. In the following section, three such missions are presented that help highlight certain cooperative decision and control challenges. Then, the major theoretical challenges in fielding cooperative UAV systems are discussed.

### 1.3.1 Example Mission Scenarios

**Combat Intelligence, Surveillance, and Reconnaissance**

A combat intelligence, surveillance, and reconnaissance (ISR) scenario is shown in Fig. 1.8 [3]. The objective in this scenario is to track moving ground targets with sufficient accuracy to enable a weapon, guided by the global positioning system (GPS) and receiving continuous target position and velocity updates, to strike the target. There is at least one stand-off sensor platform, one or more stand-in sensor platforms, and a stand-in weapon platform. Stand-off platforms are those that can perform their mission while flying a long distance away from the target area. Stand-in platforms are those that must operate in the vicinity of the target area. The stand-off sensor platform has a ground moving target indication (GMTI) Doppler radar that can detect moving targets in a large field of view when the range rate is over a certain threshold (e.g., 15 mph).

The GMTI radar has a significant target positional error ellipse, which is much greater in azimuth than in range, and therefore is not accurate enough for GPS-guided weapons. The stand-in GMTI platform has a similar target positional error ellipse. If the angle between the two views is large enough, for example, greater than 45°, then the resultant computed error ellipse from the two views may be sufficiently accurate for GPS-guided weapons. For servicing multiple targets, there is a sequence of joint tasks. This is a complex multiple-vehicle cooperative decision and control problem that includes trajectory generation, task assignment, task tours, task precedence with timing constraints, and task duration constraints. This problem, investigated in [3, 9], will be discussed in depth in Chapter 5.

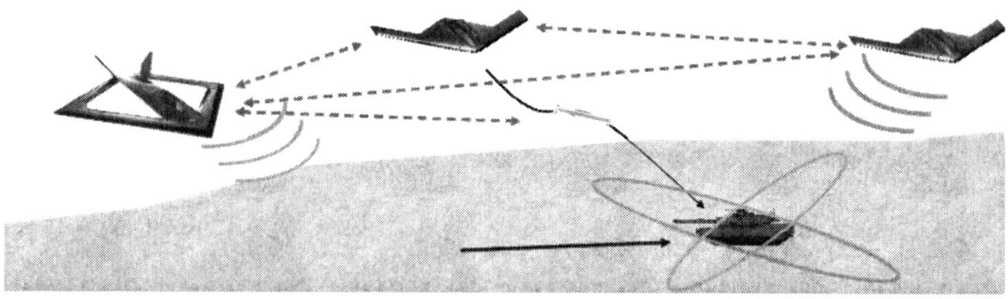

**Figure 1.8:** Cooperative ground moving target attack, combat ISR.

### Electronic Attack

The scenario for deceiving a network of radars by cooperatively generating a phantom track is shown in Fig. 1.9 [6]. To generate a valid track there is a need to assign an electronic combat aerial vehicle (ECAV) to each radar. The ECAVs, which are designed to be virtually invisible to the radars, use range delay techniques on the return radar signal, which the radar interprets as a target at a much greater range than the ECAV. The circles represent victim radars that are networked together so that tracks can be correlated. The filled triangles represent the positions of the ECAVs at time $t(1+n)$, and the unfilled triangles represent the positions of the ECAVs at time $t(1)$. The solid curved arc represents the phantom track. All the individual phantom tracks generated by the ECAVs must agree for any of the tracks to be considered valid by the network. For the tracks to agree, the ECAVs must all be simultaneously on their respective line of sight (LOS) from the radars to the phantom target. The evolution over time of the intersection point constitutes the phantom track. The cooperative decision and control problem is to coordinate the trajectories of the ECAVs to generate the required track while satisfying the vehicles' dynamics and constraints. This problem was investigated in [6, 10].

### Urban Operations

A primary application for small, mini, and micro air vehicles is tracking and positive identification of targets in a cluttered urban terrain with obstacles. Micro aerial vehicles (MAVs), in particular, can get close enough to the target undetected to allow an image of sufficient detail to be taken so that an operator can perform a high-confidence classification [5]. In addition, a MAV can achieve viewing angles around obstacles that are denied to a higher-flying asset. For continuous surveillance, a MAV is severely limited in range and endurance.

A small and micro UAV team scenario is shown in Fig. 1.10, where the small vehicle carries several MAVs to the area of operations. The small vehicle has a sensor yielding an overview of the entire area but does not have the resolution or aspect angle for positive identification. The small vehicle selects objects of interest in the field of regard and computes a feasible trajectory for the MAV to fly around obstacles and pass over these objects (potential targets). The MAV video is relayed to the small UAV for automated queuing; it is then sent to the operator for target recognition. The cueing function is selectable by the operator. The cooperative decision and control problem is to determine the assignment and paths of

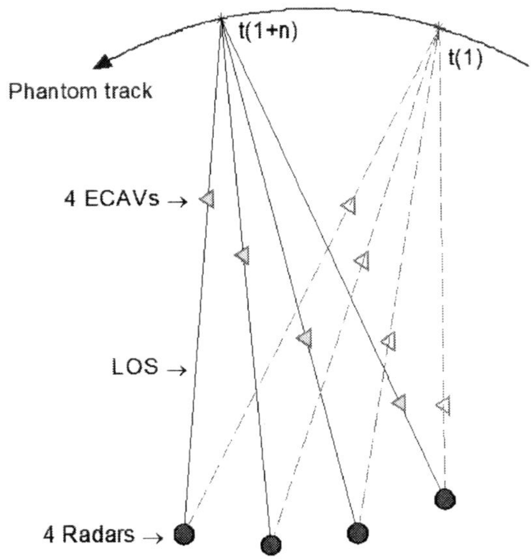


**Figure 1.9:** Electronic attack diagram.

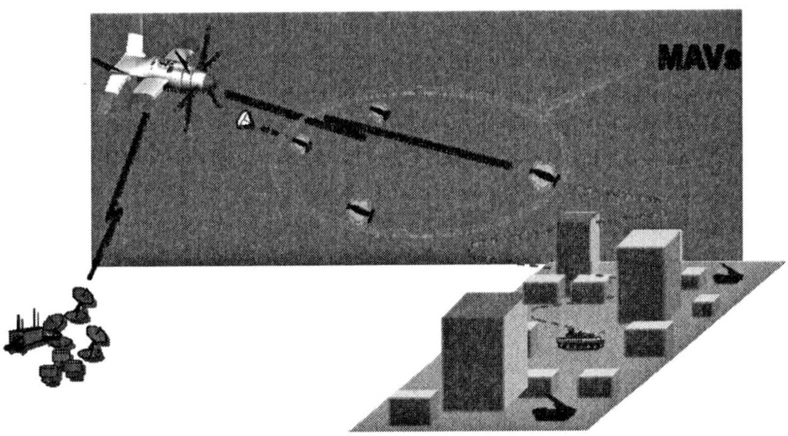


**Figure 1.10:** Urban operations scenario.

MAVs to objects of interest while taking into account that each MAV might be required to visit multiple objects with limited endurance. Such a problem was investigated in [11].

### 1.3.2 Attributes

The main attributes associated with cooperative decision and control problems, in scenarios such as the ones described above, are presented next.

### Complexity

Cooperative decision and control problems are complex. The sheer size of the problem (e.g., number of vehicles, targets, and threats) is one form of complexity. However, in scenarios such as combat ISR, coupling between the completion of the different tasks and coupling between the assignment process and trajectory optimization have the most significant impact on complexity. For example, if the vehicles each have a default task of searching, then performing it cooperatively introduces extensive coupling in their search trajectories. Once a target has been found it may need to be simultaneously tracked by at least two vehicles and attacked by a third, which further imposes coupling between the trajectories of different team members.

### Imperfect Information

Most cooperative teams will operate with limited information. The full information state is assumed not to be available anywhere in the network. The challenge is to perform the tasks cooperatively to achieve mission objectives under limited information. A specified level of team performance and team coherence requires a minimum level of shared information. This should include the team objective function, a subset of the events, and a set of functions that represent the ability of the vehicles to perform actions. Ideally, there should be sufficient information to ensure that all the tasks are covered and the tasks are consistent. Also, the vehicles may have state estimation and prediction models to provide information that is not readily available. Information flow imperfections, such as communication delays, may produce different information sets for the different vehicles in the group, which can lead to multiple strategies. The result can be uncoordinated assignments, such as multiple vehicles wrongly assigned to perform the same task on a certain target, while leaving other tasks unassigned.

### Implementability

Cooperative decision and control algorithms must take into account many factors to be fielded on UAVs in real-life missions. The number and type of factors depend on the UAVs and the mission, but a couple of the predominant factors are real-time operations and flyable trajectories.

For real-time operations, the time available to make cooperative control decisions is determined by the mission parameters. For instance, given that a mission requires that high-value targets be prosecuted immediately upon discovery, then because of travel time, if any of the UAVs are close to the target there is little time to make an assignment decision. The ability to make decisions in the required time frame is a function of the cooperative control algorithms and the capability of the data processors on the UAVs. While finding optimal cooperative decision and control solutions can be computationally intractable, it is possible to implement suboptimal algorithms that calculate solutions quickly and that can then be improved over the available decision time window. Many of these suboptimal algorithms use Euclidean distances between targets and vehicles to calculate assignment costs and timing constraints. Combining the straight-line trajectories into a multiple task assignment can result in commanded trajectories that cannot be flown by fixed-wing UAVs having a minimum turning radius. This can result in assigning tasks that cannot be executed and could lead to degradation of team cooperation.

## 1.4 Summary

This chapter laid the foundation for the presentation of cooperative decision and control algorithms in the remainder of the book. First, a brief overview of the evolution of unmanned systems, from the early use of single remotely operated balloon bombs to cooperative teams of sophisticated autonomous UAVs, has been presented. Then the concept of system autonomy has been discussed and the motivation for autonomous team cooperation, stemming from the expected synergy, has been highlighted. Next, different classes of cooperative decision and control problems and a few examples have been presented. Finally, the attributes associated with such problems have been introduced.

In Chapter 2 a more detailed discussion of cooperative teams is given; along with the associated challenges, and the trade space of the algorithms that were investigated as part of this research. In Chapter 3, the assignment of simply coupled tasks is presented. In Chapter 4 coupled tours are addressed using several different algorithms. In Chapter 5 joint tasks with time windows are introduced. Chapter 6 covers the estimation used to address asynchronous data states between vehicles. Chapter 7 address the key issue of false information, particularly false alarms.

The emphasis in this book is on coupled tasks, primarily because the militarily relevant scenarios we have investigated were dominated by coupling. In addition, many of the solutions tend more to centralized (though not full-information) than to decentralized control due to the difficulty of enforcing the dominant coupling constraints. Major sources of uncertainty addressed in this book are different data states and false information.

There are numerous topics that are not addressed in this book. Including the operator as part of the team has been a significant research focus, but is covered in a separate document. Decentralized control is not emphasized due to the extensive task coupling. While programmed responses of targets is addressed, the actions of intelligent adversaries is not addressed due to the intractable complexity these game-like scenarios introduce. Some communication issues are addressed, but range limits are not. This is because the problems investigated have limited team sizes that are not highly spatially diversified.

## Bibliography


[1] Balloon Bomb. More about balloons. *Scientific American*, 4(26):205, 1849.

[2] B. T. Clough. Metrics, schmetrics! How the heck do you determine a UAV's autonomy anyway. In *Proceedings of the Performance Metrics for Intelligent Systems Workshop*, Gaithersburg, MD, 2002.

[3] D. B. Kingston and C. J. Schumacher. Time-dependent cooperative assignment. In *Proceedings of the American Control Conference*, Portland, OR, 2005.

[4] K. Munson, editor. *Jane's Unmanned Aerial Vehicles and Targets*, vol. 14. Jane's, 2000.

[5] M. W. Orr, S. J. Rasmussen, E. D. Karni, and W. B. Blake. Framework for developing and evaluating MAV control algorithms in a realistic urban setting. In *Proceedings of the American Control Conference*, Portland, OR, 2005.

[6] M. Pachter, P. R. Chandler, R. A. Larson, and K. B. Purvis. Concepts for generating coherent radar phantom tracks using cooperating vehicles. In *Proceedings of the AIAA Guidance, Navigation, and Control Conference*, Providence, RI, 2004.

[7] C. M. Peebles. *Dark Eagles: A History of Top Secret U.S. Aircraft Programs*. Presidio Press, Novato, CA, 1999.

[8] A. C. Piccirillo. The origins of the anti-ship guided missile. In *Proceedings of the 1997 World Aviation Congress Conference*, Anaheim, CA, 1997.

[9] T. Shima and C. Schumacher. Assigning cooperating UAVs to simultaneous tasks on consecutive targets using genetic algorithms. *Journal of the Operational Research Society*, 2008.

[10] T. Shima, P. Chandler, and M. Pachter. Decentralized estimation for cooperative phantom track generation. In *Theory and Algorithms for Cooperative Systems*, Series Lecture Notes in Economics and Mathematical Systems 58, D. Grundel, R. Murphey, P. Pardalos, and O. Prokopyev, eds., Springer, New York, 2007.

[11] T. Shima, S. Rasmussen, and D. Gross. Assigning micro UAVs to task tours in an urban terrain. *IEEE Transactions on Control System Technology (special issue on Multi-Vehicle Systems Cooperative Control with Application)*, 15(4):601–612, 2007.

[12] K. P. Werrell. *The Evolution of the Cruise Missile*. Air University Press, Washington, DC, 1985.

# Chapter 2
# Challenges

***Phillip Chandler and Meir Pachter***

*Nothing is more difficult, and therefore more precious, than to be able to decide.*
—Napoleon Bonaparte

In this chapter the challenges of cooperative control are examined. While the dynamics of unmanned air vehicles are important, considerations relevant to resource allocation and optimization, the prevailing information pattern, and unstructured environment/uncertainty, dominate the analysis. The potential benefits of cooperative team actions and, in general, the benefits of cooperation among distributed controlled objects are addressed.

## 2.1 Introduction

This chapter is not a survey of all that has been done in cooperative control, but rather it is a compilation of the work by a group of researchers, with special emphasis on military applications. Nor is this chapter intended as a comprehensive analysis of cooperative control applications, but rather focuses on a streamlined surveillance scenario. A team consisting of a flight of UAVs is tasked with searching for and recognizing multiple targets in a battle space with many false targets. Allowing for an unstructured environment in cooperative control operations is essential. The presence of false targets and clutter forces one to consider a probabilistic setting. While the inner-loop dynamics of unmanned air vehicles are important, and, in particular, the dynamics and aerodynamics of small air vehicles pose new research issues, the challenges relevant to task coupling, partial information, and uncertainty dominate the analysis. Given these challenges, the potential benefits of cooperation among distributed agents are addressed in this chapter.

## 2.2 A Taxonomy of Teams

A team is here defined as a loose collection of spatially distributed controlled objects, a.k.a. UAVs, that have some objectives in common [26, 34]. Air vehicles may be too restrictive a term—generically, a team consists of members, or agents, and the team can (and generally will) include humans as operators, task performers (think of target recognition), and/or supervisors. The presence of a common objective forges a team and induces cooperative behavior. If the air vehicles are working together to achieve a common objective, then they are considered a team. Different degrees of team interaction are possible: coordinated, cooperative, and collaborative. At the same time, additional individual objectives of the team members can encourage team members to opt for a weak degree of noncooperation, mild competition, or outright adversarial action.

When team decision and control problems are discussed, it is important to address the unstructured environment/uncertainty, the organizational structure, the information pattern, and task coupling. Individual operational scenarios can be dominated by one of the above but will contain elements of all. The interaction of these different facets of a team problem cannot be ignored. The vehicles are coupled through the performance/objective function and the task constraints. The objective function is discussed in this section. The task constraints are discussed in later chapters. In the following, three degrees of team interaction that stem from the coupling in the objective function are discussed.

### 2.2.1 Team Coordination

Team coordination is the strongest degree of cohesive team action. Consider a set of UAVs that have been designated to be part of the team; all share a single team objective and thus strive to optimize a single payoff function. The team could have more than one payoff function that it wishes to optimize, which would then entail multiobjective optimization [33]. Often, the different payoff functions can be assigned weights and rigorously combined into a single objective function. There is no conflict of interest among the team members; otherwise an incentive scheme [24] would need to be devised. Conflict of interest is defined here as the UAVs having different objective functions, which is discussed in detail in the next section. The important distinction in team coordination is that particular team members do not have individual objective functions: a team member is a resource that can be overutilized or underutilized, if that will best achieve the team objective. The team members are obligated to participate and any assignments, tasking, or agreements are binding; they cannot opt out. At the same time, the forming of coalitions is not possible. The team may be geographically distributed, but it operates as a single unit.

### 2.2.2 Team Cooperation

Each of the team members has a private objective function which he strives to optimize, in addition to the team objective function. The private and team objective functions are weighted subject to $0 \leq w \leq 1$. A weight of $w = 1$ on the private objective function means the member acts in its own self-interest, in which case there is no team action. A range of $0 \leq w \leq 1$ on the private objective functions corresponds to an increasing level of cooperation of the team members, to where $w = 0$ entails strict coordination. There is a possibility for conflict of interest, but, due to the structure of the objective functions used, it is not generally dominant. In most cases, local objectives such as fuel conservation and survivability are not in conflict with the team objectives and can be jointly achieved. Thus, the multicriteria optimization aspect of the problem is not dominant and a weighted sum of the objective functions yields a conventional optimization. If they are in conflict, the team objective takes precedence according to the weight used.

### 2.2.3 Team Collaboration

Team collaboration is a loose form of team interaction. In some cases this can be the result of the task assignment or resource allocation method used [6, 8, 9, 11, 25]. At the global (team) level, the focus is on task completion, that is, feasibility. Each team member tries to maximize his local objective function consistent with team task completion while avoiding tasking conflicts. This requires that a protocol be designed for negotiation to arbitrate conflicts [31, 41]; this connects with the currently developed theory of communication networks. Ideally, those tasks that are best for each member to perform according to their private objective function are tasks that need to be done for the team objective function. In addition, there is the implicit constraint that the selected tasks are not redundant. Here there are possibilities of significant conflicts of interest: if the team members have a set of equally valuable (to them) tasks, then likely the conflicts can be resolved (mutually agreeably), and collaboration can be as efficient as coordination. Obviously, the more tightly coupled the various team tasks are, the more difficult it is to achieve a collaborative solution. Strong coupling will occur if a homogeneous team of multirole UAVs is employed, or if the battle space is small. A coordinated or cooperative operations approach will be needed. Also, negotiation is not compulsory; the members are not forced to participate. If a solution for a particular team member cannot be found, then this team member can opt out and join an alternative team that has a better overall match of member objective to team objective.

### 2.2.4 Goal-Seeking Team Action

Goal-seeking team action is a further abstraction of a team decision and control problem. Here there are no a priori designated team members. The team is formed from a set of available resources that are loosely networked. Each UAV can simultaneously be a member of several teams. Once a team's objective is achieved, the team will dissolve. The goal in general is abstract and has to be converted to a sequence of intermediate objectives or milestones, and the objectives in turn have to be converted into a set or sequence of tasks that are assigned to, and coordinated between, the team members. There might be an intrinsic conflict of interest between the goals of the teams that the UAV can simultaneously be a member of. The teams are therefore self-organizing [27]. This results in a high level of autonomy, where the vehicles are independent agents that however work together in an ad hoc fashion, as needed, to achieve an overarching goal. Each vehicle also strives to optimize its utility function, which may be handled through coordination, as previously mentioned.

### 2.2.5 Noncooperative Behavior

To this point we have been describing the different modes of how UAVs interact with other UAVs in a team. In a team, by construction, the objectives are basically compatible.

We now address the control of UAVs that are operating in teams that are competing and, possibly, adversarial. If there are two teams, this is the domain of a significant part of game theory research. This includes strictly competitive zero sum games and also non zero sum games, e.g., bimatrix games [33, 63]. This is the field of much military research, as in the war game of Blue team versus Red team. However, the field is much richer than this, because in reality there can also be, for example, a White team and a Green team. What's more, membership in a team can change fluidly, and the teams can form coalitions which can also dissolve. This is a rich area in which complementary and conflicting interests, collusion, hidden objectives, signaling, diversion, gambits, propaganda, and disinformation play a significant role. Conflict of interest is discussed in the next section.

## 2.3 Conflict of Interest

Conflict of interest is the signature characteristic of noncooperative behavior. Conflict of interest is brought about when the UAVs have different objective functions which, in addition, are structured such that joint action which simultaneously optimizes the different objective functions is not possible or feasible. Thus, consider the performance functionals $J_1(u, v)$ and $J_2(u, v)$ for UAV 1 and UAV 2, respectively; the decision variables of UAV 1 and UAV 2 are $u$ and $v$, respectively. The payoff for UAV 1 is a function of the actions (or decisions) $u$ taken by UAV 1, as well as the actions (or decisions) $v$ taken by UAV 2. Each affects the value of the other's objective function. If the objective functions were not coupled, that is, $J_1(u)$ and $J_2(v)$, then obviously there is no conflict of interest. Alternatively, if $J_1(u, v) = J_2(u, v)$, there also is no conflict of interest (team objective).

If this is not the case and a plan communicated ahead of time (an agreement) is not in place, then this definitely is a noncooperative scenario. The question is, What strategy does UAV 1 use to determine the "best" action $u$ to minimize his cost $J_1$? Similarly, UAV 2 is faced with the problem of minimizing his cost $J_2$. The actions $u^*$, $v^*$ are "best" if $J_1(u^*, v^*) \geq J_1(u^*, v)\forall v$ and $J_2(u^*, v^*) \geq J_2(u, v^*)\forall u$. This means that if UAV 2 deviates from $v^*$, then UAV 1 will do better than $J_1(u^*, v^*)$, and if UAV 1 deviates from $u^*$, then UAV 2 will do better than $J_2(u^*, v^*)$. Thus, $J_1(u^*, v^*)$ and $J_2(u^*, v^*)$ constitute guarantees for the respective players, no matter what the other player or vehicle does. This is a Nash (noncooperative) equilibrium point [33, 63].

Now consider the definition of "best" where $u^*$ and $v^*$ is the point where no further improvement (reduction) in $J_1$ can be made without an increase in $J_2$, and conversely, no further improvement (reduction) in $J_2$ can be made without an increase in $J_1$. This is a Pareto (cooperative) equilibrium point [33].

There are two problems associated with the Pareto equilibrium concept.

1. Should UAV 1 choose another action $u'$ and deviate from $u^*$ whereas UAV 2 sticks to his Pareto choice of $v^*$, then $J_2(u', v^*) > J_2(u^*, v^*)$ and $J_1(u', v^*) < J_1(u^*, v^*)$. Now UAV 1 did better, at the expense of UAV 2.

2. In general, Pareto equilibria are not unique and therefore an agreement is needed or side conditions imposed on the actual Pareto equilibrium used.

If both vehicles cooperate and agree to play the Pareto equilibrium solution, then they both can do better than the outcome provided by the Nash equilibrium. This is the benefit of cooperation. However, the "agreement" to play Pareto must be rigidly enforced; otherwise one side can choose an action that results in a one-sided benefit to one of the teams, at the expense of the other. If the "agreement" cannot be rigidly enforced, then the players are better off playing Nash, because at least they'll get the guarantee. The latter can be computed ahead of time, before the game is ever played. One might then decide whether it's at all worth playing the game.

In the context of human behavior, Pareto games provide an incentive to cheat [18]. Hence, the "contract" must specify a penalty for one of the parties breaking it. If the parties are strictly self-interested, an expected cheating value calculation can be made which is a function of [Reward $-$ Penalty $*$ P(caught|cheat)]. Of course, the other party in the agreement can make the same calculation and both could violate the agreement, which means that both parties could end up with less than the Nash (noncooperative) value. This much-studied predicament is referred to as the "prisoners' dilemma" [33].

It is not at all clear if these considerations have much bearing on UAV team performance, but they abound in human teams. For example, in team sports each of the members shares the objective of his team winning the game by accumulating more points than the opposing team. There are a series of plays and roles that are agreed on that could be considered a Pareto solution. However, one of the players might have a different objective that is not revealed up front. That is, his objective is to maximize the points attributed to him, not necessarily to win the game. His team mates stick to the playbook, he scores more points, he wins (becoming more valuable), and his team loses.

Concerning adversarial behavior in a UAV team, consider a network (team) of geographically distributed assets that have a range of overlapping capabilities. These assets service targets as well as provide services to other UAVs. These services have a variety of values and associated costs. Each UAV attempts to provide the highest-valued services at the lowest cost [49]. Still, if each of the vehicles, driven by its above-stated self-interest, engages in the pursuit of maximizing value and minimizing cost, then, under mild assumption and in realistic noncontrived scenarios, this should cause minimal churning and lead to maximizing the value for the team. In the context of UAV teams, situations are not envisaged where the UAVs have very different objectives (self-interest), exhibit predatory behavior toward each other, or actively practice deception. Such cases are not part of the team behavior in the following chapters.

## 2.4 Distributed Decision and Control Systems

Fig. 1.7 introduced a classification of distributed decision and control systems [12, 60]. The "centralized" quadrant represents classical centralized control [7, 17]. The complete state information from the distributed UAVs is sent to a center where the decision maker (DM) resides. No independent action is taken by the UAVs. This control concept can render optimal control action insofar as complex constraints and coupling or interactions between the vehicles can be properly accounted for and not assumed away. Algorithms used here include dynamic programming [38, 46, 54], large linear programs [17], and nonlinear programming [7]. This, in turn, causes centralized control not to scale well due to the curse of dimensionality. In addition, a centralized optimal control structure might suffer from

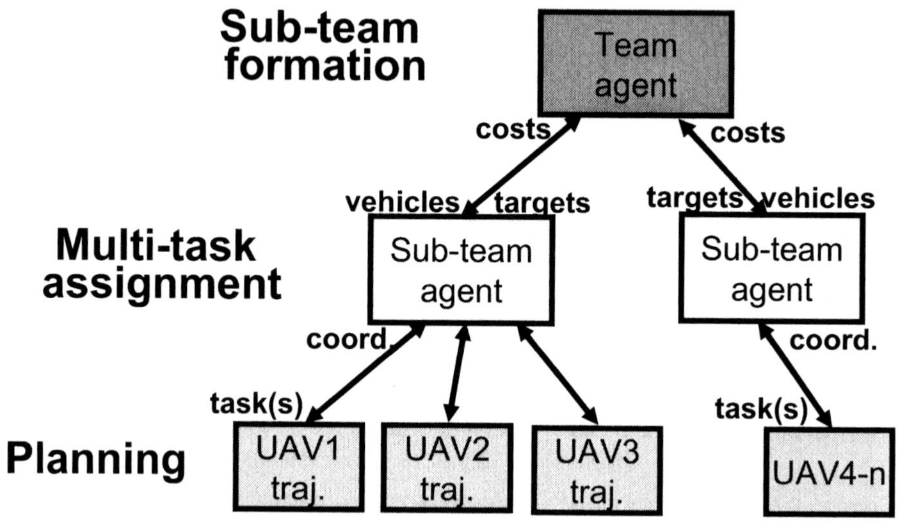


**Figure 2.1:** Hierarchical decomposition.

fragility and a lack of robustness to, e.g., missing data. Much of the following chapters will fall in this category because in general it yields high team performance.

The "hierarchy" quadrant is where decomposition is prevalent. Motivated by the structure of organizations, hierarchical control [36] schemes are used; see Fig. 2.1. The UAVs send the local vehicle state information to a higher-level DM. The team members have limited global information, but they send their individual cost functions to the higher-level DM. Consequently, an optimal assignment can be made by the DM that is beneficial to the team as a whole. While a degree of robustness is achieved, it is difficult to decompose the control problem and maintain high performance if there is appreciable task coupling. This approach is scalable, but optimality, in contrast with the centralized solution, is not achieved.

Optimization techniques used here include network flow programming [6, 9, 19, 23, 40], mixed integer linear programming [1, 52, 53, 56, 57], constraint satisfaction [37, 66], graph theoretic search algorithms [39, 50], stochastic search [59], set partition [3], the relative benefit method [48], iterative auction [6, 8, 10, 13, 29, 64], negotiation, and consensus schemes [36].

The "decentralized" quadrant represents strongly decentralized control schemes [4] which rely on minimal global information. In the limit, there could be little or no communication at all between the vehicles, and what limited partial information that is available is only inferred about the other vehicles objectives by measuring their actions (state) as seen through on-ship sensors. This line of work is often referred to as emergent behavior [27]. Often, there is an assumption that the simple low-level interactions of the vehicles will result in complex team behavior that meets a team objective. In general, however, this approach leads to a low-performance macro-level response. For example, the team maintains some loose cohesion and a nominal center of mass trajectory. The team behavior appears to be complex and one marvels at the complexity brought about by a few simple rules. However, the achieved performance here is consistently low, due to the dearth of communication and consequently little or no global information. Some techniques used here are biological analogies/biomimetics [44], mechanical analogies à la potential fields [44], parallel auction [8] which might cause churning (repeated reassignment), consensus schemes [41], and Paxos [31], which is an asynchronous voting scheme with consistency guarantees.

The "predictive" quadrant is the domain of high decentralization and, at the same time, good performance. The individual team members are highly autonomous and capable of independent action but share a common objective. Some mechanization concepts come from economics: negotiation [65], incentive games [24], consensus voting [31], and distributed constraint satisfaction [37]. Since there is little global coordinating information, there is a high dependence on state estimation and predictive models. The better these models are at estimating future states, the higher the overall performance of the team. In general, the more coupled the various tasks the vehicles are to perform, the more complex and time-consuming the arbitration to resolve the task conflicts.

## 2.5 Complexity in Cooperative Teams

In cooperative teams, an interactive decision-making process between vehicles takes place, while individual vehicle autonomy is preserved. There is a continuum between centralized and decentralized control. If a fully decentralized team means no communication, then in a cooperative team there is a requirement for the minimum level of globally communicated information that allows the desired level of team performance to be achieved.

In general, the performance of cooperative control can be characterized by task coupling, uncertainty, communications delays, and partial information. The interaction of these dimensions renders cooperative optimal control a complex problem. Currently there is no working theory of cooperative systems that takes into account all these dimensions. A hierarchical decomposition is normally tried to reduce the problem to more digestible bits, but optimality is forfeited in the process. Some degree of coupling is ignored to achieve decomposition. This results in a suboptimal solution which is traded for solvability and some degree of robustness. Many times robustness comes at the expense of optimality, and vice versa. Indeed, the optimal operating point might be sensitive to changes in the problem parameters.

Team control and optimization problems are decomposed in space, in time, or along function lines. Forming of subteams of UAVs and tasks can also be done by graph theoretic methods [42], set partition approaches [3], and relative benefit optimization techniques [48], as well as by brute force search [49]. The subteam optimization problem (see Fig. 2.1) then reduces to the multiple assignment problem: determine the task sequence and timing, for each team member, that satisfies all the constraints while minimizing an overall team objective function. The individual vehicles then perform their own task planning and send coordinating information, preferably a sufficient statistic, around the network or to a team leader. Algorithms for constrained multiple task assignment include heuristic search, e.g., branch and bound, Tabu search, or genetic algorithms [2]; generalized assignment [15]; linear programming [17]; iterative network flow [6, 23]; and iterative auction [5, 29]. One of the primary contributors to the complexity of multiple assignment is task coupling in the face of floating timing constraints—the latter brings in aspects of job shop flow optimization, or scheduling. Floating timing constraints are addressed in Chapter 5.

### 2.5.1 Task Coupling

UAV team missions such as suppression of enemy air defenses and wide area search and destroy are dominated by coupled tasks with floating timing constraints. There are a number of issues involved in solving the multiple assignment problem in a cooperative framework. Chief among these is the ability to decouple assignment from path planning for specific tasks. This means that tasks and path plans are generated to determine costs that are then used in the assignment process. The assumption is that these calculations are still valid after the assignment is made. This is even more so for tour sequences. Unless all possible tours are generated, suboptimality is being ignored when chaining together tasks. Decoupling assignment from path planning is addressed in Chapter 4.

Also, decoupling assignment from timing is often done. For example, task tours are assigned first. Then the task order and precedence are enforced. This can be done myopically to set the task time using the earliest task that needs to be done, then the next, etc.; or the task times can be negotiated between the vehicles until a set of task times is arrived at that satisfies all the timing constraints. The assumption here is that these task times will not have a different, closer-to-optimal assignment. The decomposition assumptions to address coupling may lead to infeasibility, where all the tasks cannot be assigned, as well as a significant degree of suboptimality, that is, poor performance. If the task coupling is strong, decentralization is a bad idea [49]—optimality is sacrificed, the algorithm might induce churning, and, worse, feasibility is not enforced. Iterative task assignment for precedence constraints is addressed in Chapter 3.

### 2.5.2 Uncertainty

Some cooperative team problems can be dominated by uncertainty rather than by task coupling. This is true for those missions where the target identification, target localization, threat identification, and threat location are not known in advance. Some of this information may be known, while the rest is estimated using a priori given probability distributions. The challenge is to calculate the expected future value of a decision or action taken now. For example, if the UAVs use their resources on targets now, there may be no reserves left for targets that are found later and that have higher value [20, 22, 54]. At the same time, actions taken now might decrease the level of uncertainty. The latter can be gauged using information theoretic concepts. Possible choices are to myopically follow the decision path of least risk or to follow the decision path that maximizes the possible options in the future. Of course, the safest and middle-of-the-road decisions are not generally the best. Furthermore, one critical source of uncertainty is associated with the actions of an adversary in response to an action taken by the UAVs. Possible approaches to account for uncertainty are stochastic dynamic programming [46, 54], Markov decision processes [32, 47], Bayesian belief networks [28], information theory [21], and, in the case of no information, game theory [33, 63]. Uncertainty in target location and classification with finite resources is addressed in Chapter 7.

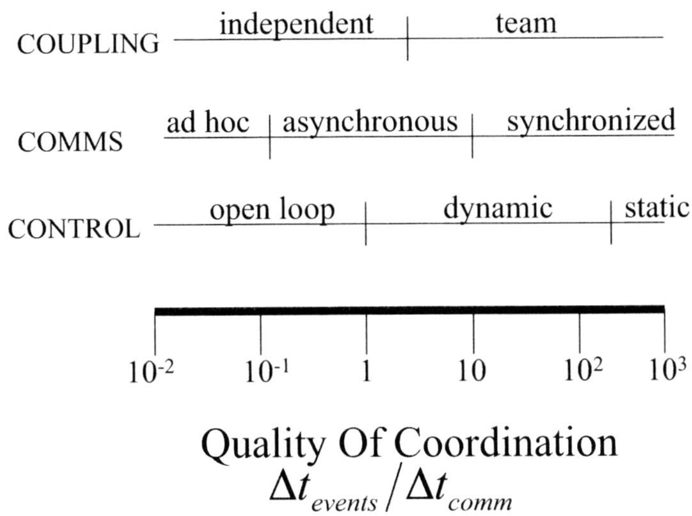


**Figure 2.2:** Notional coordination metric.

### 2.5.3 Communication

The basic premise of cooperative control is that the UAVs can communicate whenever and as much as they need to. All networks incur link delays, and if these delays are sufficiently long compared to the time between events (see Fig. 2.2), they can completely nullify the benefits of team cooperation (cooperative control). Recall that control system delays in the feedback path are conducive to instability. A critical choice here is whether the team decisions are synchronous or asynchronous. Synchronous implies that there is a common (and up-to-date) database accessible to all the vehicles. If an event occurs locally, the event and all associated information is shared across the network, and a decision based on the new event cannot occur until this happens. Under this protocol, a single actor can slow down the whole team and compromise time critical tasks. Strategies are needed to maintain team coherence and performance. Synchronous team protocols are frequently used in conjunction with a redundant centralized decision approach, where each member of the team solves the same decision problem for the whole team, thus ensuring centralized optimality. Redundant centralized is the decision system approach generally pursued in this book.

Asynchronous decision protocols, however, while more robust to delays, are much more difficult to verify to prove correct operation. They are susceptible to inconsistent information across the network, can lead to decision cycling or churning, and—worse—infeasibility. The higher the rate of occurrence of events, the more difficult these problems become because the input's frequency exceeds the system's "bandwidth." Some useful protocols include consensus voting [41], parallel computing [12, 14], load balancing [12], job shop scheduling [61], and contract nets [55, 60]. However, with false information and sufficient delays, consensus may never be reached. Indeed, false information strongly negates the benefits of cooperative control. The situation is somewhat analogous to the feedback control situation: feedback action is superior to open loop control, provided the

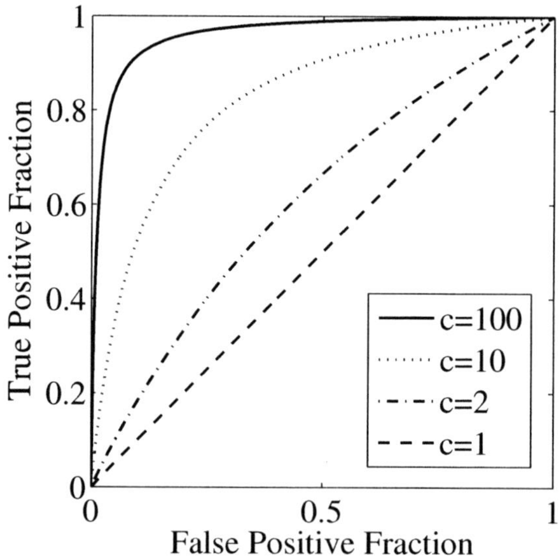


**Figure 2.3:** Family of receiver operating characteristic.

signal-to-noise ratio in the measurement is high. One then takes advantage of the benefit of feedback. If, however, the measurements are very noisy, one might be better off ignoring the measurements and instead opt for open loop (feed-forward or model-based) control.

### 2.5.4 Partial Information

Decentralized control [30] as well as cooperative teams are characterized by limited, or partial, information. The full information state is not available anywhere in the network. Worse, the information pattern is not nested. The challenge is to perform the tasks cooperatively and achieve a degree of optimality under limited and distributed information.

A specified level of team performance and team coherence requires a minimum level of shared information (e.g., the team objective function) a subset of the relevant events (e.g., pop-up target information), and part of the state vector (e.g., the fuel state of the UAVs). There should be sufficient information to ensure that all the tasks are accounted for and the tasks are consistent. This can be considered the "sufficient statistic." Also, the vehicles may have state estimation and prediction models to provide information that is not available locally. Interestingly, the vehicles may have different objective functions as well as inconsistent and delayed information that can result in conflicts that need arbitration or negotiation. False information (see Fig. 2.3), in particular, can induce poor performance in a cooperative team—one may be better off using noncooperative control.

For a fixed parameter $c$ in Fig. 2.3, the operating point on the receiver operating characteristic (ROC) curve determines both the probability of correct target classification and the probability of a false positive. The parameter is set by the flight condition, and the operating point is determined by the threshold setting in the sensor. False information is addressed in Chapter 7.

### 2.5.5 Operator

Fundamental to the field of cooperative control is the level of realized team autonomy. A continuum of team autonomy levels is possible, from completely autonomous action from release, to human operator management by exception, to human operator management by permission—the latter being the least autonomous (also called mixed initiative). These different autonomy levels can exist as a function of engagement phase or current task and can dynamically change as a function of system state. Furthermore, an autonomous team is a network of distributed functionality, where, for example, an operator could provide the critical target recognition functionality (difficult to accomplish by machine), interface to outside systems, or provide supervisor functions. Autonomous target recognition is a long-range goal and it therefore makes sense to assign the object classification task to a human operator.

In the context of stochastic cooperative control, the operator can introduce significant uncertainty. Chief among these are operator errors, e.g., object classification errors, and delays. If the team consists of a number of vehicles all sending sensor streams to an operator, the operator's workload can be high, which would increase the operator's delay and error rate. For time-critical UAV operations, a high degree of automation of team decision and control is required. To this end, a model of the human operator's performance is needed [45], e.g., the probability distribution function of the operator delay or the operator's error rate—as in a Poisson process model [43].

Close coordination of a UAV team, including the operator, can be maintained despite significant communication delays and processing delays associated with operator cognition phenomenology, operator workload, and operator errors. These operator characteristics must be captured in a stochastic model and incorporated into a controller obtained by solving a stochastic program. This enables robust team coordination to be maintained despite a significant level of uncertainty brought about by operator misclassification of objects of interest, as well as delays. Operator issues are not emphasized in the later chapters of this book but are addressed in ongoing work by the research group.

### 2.5.6 Adversary Action

Much of the research done to date on cooperative teams has been with passive targets—not allowing for intelligent adversary action.

Although threats were incorporated into simulation models, the targets did not react to actions taken by the UAV team in the work described herein. While in some simulation studies threats do react to UAV actions, behavioral/reactive modeling is routinely used, namely, the threats' reactions are preprogrammed and the strategy is known ahead of time to the UAV team. For example, should a UAV penetrate a missile engagement zone, it will be fired upon. For the known missile emplacements, UAV trajectories are planned around the zone ahead of time. If there is an approximate distribution of threats, then UAV trajectories are planned based on the threat expected locations. Unplanned for threats encountered by the UAV trigger a fixed set of UAV responses, such as evasion, or a deviation around the new threat is replanned. Addressing adversary action on the fly is fundamentally different in that the intelligent adversary observes the state of the engagement, is cognizant of the attacker's capabilities, and solves for his optimal strategy. Here a UAV or team of UAVs must derive a strategy based on the possible actions the adversary may take for every action that the UAV team may take. This two-sided optimization problem explodes exponentially for increasing sequences of action-reaction moves. This can be cast as a dynamic programming problem if all the possible action-reaction pairs are known. Such two-sided optimization problems are solvable offline for a modest number of stages, that is, a few moves deep, so that the computed optimal control strategy can be implemented online.

Making the situation even more complex is the possibility of adversary deception such as the employment of decoys, diversions, traps, and false information. For example, the employment of decoys complicates the operator's classification task and thus causes an increase in the rate of false positives. This, in turn, depletes the team's resources, adversely affecting the team's performance. Adversary actions are not addressed in this book.

## 2.6 Algorithms for Cooperative Control

Following is a partial list of algorithms that could be used for cooperative control of a team of UAVs; it is not intended as an survey. We have investigated the following approaches: a relative benefit optimization method [48], network flow [40], iterative network flow [23, 49], iterative auction [5, 6], decomposition (assign, then time ) [49], parallel auction [64], combinatorial or bundle auction [25], mixed integer linear programming (MILP) [52, 53, 56, 57], partially observable Markov decision processes (POMDP) [32], Bayesian belief networks [28], genetic algorithm or generalized search [44, 49, 58, 59], centralized or distributed constraint satisfaction or optimization [66], tree search [50], stochastic dynamic programming [46, 54], job shop scheduling [61], vehicle routing [51, 62], parallel computing [12], voting or consensus [31, 41], contract nets [60], games [33, 63], receding horizon to periodically reevaluate the strategy [16, 35], and multiple agent systems [65].

While not exhaustive, this is a good cross section of the available options for cooperative control algorithms synthesis. We have found that for strong task coupling, the centralized algorithms such as MILP and dynamic programming can give optimal solutions for problems that are computationally tractable. For scenarios with weak task coupling, network flow and auction protocols are reasonable approaches, while for scenarios with an intermediate level of task coupling one can call on iterative methods, including relative benefit. Many forms of uncertainty can be accounted for using stochastic dynamic programming, POMDP, or Bayesian belief networks. Some of these can also account for various levels of coupling, but they are not readily decomposable. More strongly decentralized approaches such as distributed constraint satisfaction and parallel auction cannot easily capture coupling among different tasks. Asynchronous parallel computing may address some of the communication issues previously discussed.

In summary, there is no one approach that addresses all the manifold facets of cooperative control. The development of heuristic methods and iterative schemes geared to addressing the dominant traits and the specifics of a particular operational scenario is required. The remaining chapters relate our experience with some of the approaches we found most suitable.

### Coupling of Tasks versus Decentralization versus Communication

Each of the distributed decision and control approaches or algorithms discussed, as applied to the cooperative team problem, has certain advantages which can be used in the team

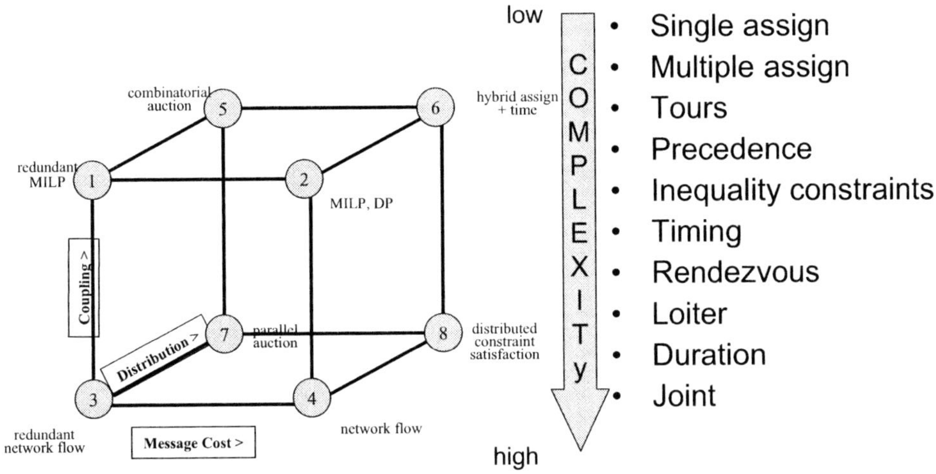


**Figure 2.4:** Coupling–decentralization–communication trade space.

problem at hand. A notional trade space is shown in Fig. 2.4, where strength of task coupling, communications volume/bandwidth, and the degree of decentralization feature. The corners of the cube, labeled 1 through 8, show those approaches that appear to be more suitable to address the problem characteristics represented by that node.

The origin, or node 3, is typified by weak task coupling or few constraints; full information, that is, a centralized or nested information pattern; and very low communication costs—high bandwidth, near-zero delays, and complete connectivity. The single assignment optimization problem with binary decision variables and simple constraints can be posed as a network flow problem. This class of integer programs can be solved using slow relaxation techniques or fast specialized binary tree search algorithms. The control can be partially distributed through implicit team coordination (redundant centralized), which is where the centralized solution is computed, redundantly, by each of the UAVs. Network flow is addressed in detail in Chapter 3.

As the cost of sending messages increases, the need for redundancy could motivate us to use the centralized network flow approach shown at node 4. Concerning node 3, as the tasks become more strongly coupled and, particularly, when floating timing constraints are included, the network flow solution is not suitable. MILP can rigorously include both integer and continuous constraints. With a full information (centralized) scheme, the same approach can be used to have the solution computed redundantly by each of the vehicles, as at node 1. As the cost of sending messages increases, the need for a degree of redundancy leads one to use the centralized MILP algorithm, which could also be formulated as a dynamic program at node 2. Receding horizon optimization can reduce the computational burden, enhancing scalability, and also reduce the need for information about the environment, but at the expense of optimality. MILP is addressed in Chapters 4 and 5.

One objective of cooperative teams is to distribute the control so that the UAVs can operate more autonomously but cooperate in performing team tasks. Decentralization increases in moving from node 3 to node 7. The Jacobi auction [29] iterative algorithm, while intuitively appealing, is, in its basic form, mathematically equivalent to the network flow optimization problem. This iterative optimization scheme does not require full information locally, but it does require a centralized auctioneer with an attendant increase in the message load. Continuing to node 7, the centralized auctioneer is eliminated in the asynchronous parallel auction scheme. Again, this is feasible because only the simple constraints (coupling) of network flow are active, and the very low message cost does not incur a penalty on iteration (bidding). The full benefits of parallel computing are not achievable, since targets cannot be processors. Network flow as a noniterative form of auction is addressed in Chapter 3.

As the task coupling strength increases at node 7, the simple constraints, easily accommodated using parallel auction-based algorithms, are no longer appropriate. Combinatorial auction at node 5, where bundles of goods are traded, is a mechanism by which more extensive task constraints can be included: coupled tasks are bundled together during the UAV bidding phase. In the worst case, every feasible permutation could be a separate object to bid on. Preparing such bids may require full information. Timing constraints cannot be readily incorporated, and combinatorial auctions cannot be directly layered on top of a parallel auction. From node 7, as communication costs increase, parallel auction-based protocols are no longer appropriate, since the latter are contingent on near instant messages. Distributed constraint satisfaction type resource allocation algorithms at node 8 may be preferable because fewer messages are needed, and they are strongly decentralized. Only simple constraints can be accommodated here, however.

Finally, at node 6, we have the most challenging case of strong coupling, minimal communications, and strong decentralization. No one approach or solution could address these requirements, and, in some cases, no suitable approach currently exists which addresses all the requirements, since, in some sense, they are contradictory or mutually exclusive. One practical strategy is to apply a network flow, auction, or parallel auction algorithm iteratively for a fixed (receding) horizon. While not optimal, this allows more complex constraints to be incorporated, including timing constraints. This is a two-stage strategy, which partially decouples task assignment from task planning. The minimum deviation task times allow the vehicles to refine their trajectories (or bids) to meet the team task constraints. This hybrid approach is addressed in Chapter 3.

## 2.7 Observations

Fig. 2.5 shows that the work on cooperative control draws from three established disciplines: control, operations research (OR), and computer science (CS), as well as elements of many other disciplines, including estimation, statistics, and economics theory. The research challenge has been to combine these disciplines to form the beginnings of the new integrated discipline of cooperative control. Following is a list of observations made over the course of the research effort that will provide some insight into what has been done and what yet needs to be done.

- A comprehensive theory of cooperative control must include uncertainty, communication costs, the consideration of local versus global information structures, nested versus nonnested information patterns, control decentralization, task coupling, predictive models, adversary action, false information/false targets and false positives, and well-reasoned performance measures.

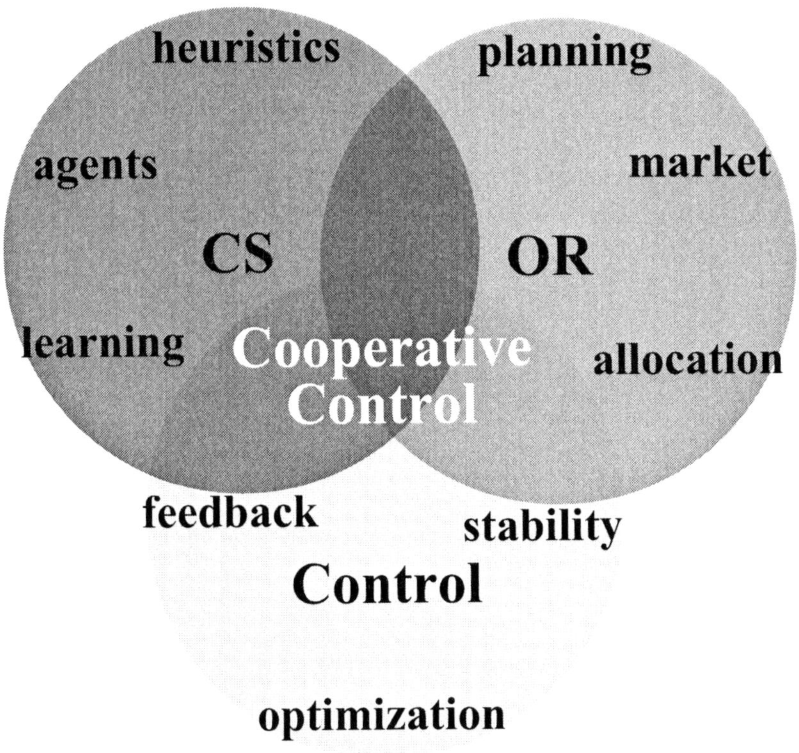


**Figure 2.5:** Cooperation is at the intersection of disciplines.

- It is not always beneficial to cooperate, particularly if there are long delays or if the sensed or computed information that is communicated is very noisy or error prone.
- It is possible to obtain the optimal solution only for moderately complex operational scenarios.
- Decomposition is a necessary attack upon complexity, but guarantees of optimality are forfeited and, more important, oftentimes feasibility is not guaranteed, leading to the possibility of task churning.
- Jacobi auction-type assignment algorithms are an iterative form of network flow algorithms which yield a degree of decentralization, but at the expense of possibly much more communications.
- The evaluation of different solutions is problematic. Comparisons can be made, but all solutions may be far from the optimum. There is a need to obtain analytical solutions for small-scale problems which should help benchmark the array of heuristic "algorithms"/recipes/procedures currently developed.
- Aside from optimality, feasibility is an even more important concern. In general it will not be possible to prove that a procedure will not generate infeasible results.

- The truth might never be known because in general there are critical states that are not observable. Hence, randomization and mixed strategies are called for, as are Monte Carlo–based simulation studies.

- Due to sensitivities to random disturbances, adversary action, operator performance characterization provided in the form of statistical data, and randomized strategies, extensive simulation studies are required for objectively evaluating competing cooperative control schemes.

- As per the employment of feedback, the benefits of cooperative control are questionable when measurement noise and bad or delayed information are dominant factors. Networks are especially good at rapidly spreading bad information.

- False target attack rate dominates the wide area search and destroy scenario.

- Highly reliable target recognition is *the* critical capability to make autonomous attack vehicles a reality.

- In general, a strongly decentralized controller cannot recover centralized controller performance except if the tasks are nearly independent, that is, the optimization problem at hand is virtually decoupled—a rather trivial cooperative control problem.

- For highly coupled tasks, a strongly decentralized controller will need vastly more messages than a centralized controller due to the need to constantly resolve conflicts. This introduces a degree of vulnerability.

- State estimation is the central actor in addressing partial information. In the absence of observability, the illusion is created of being able to provide state estimates from recursive Kalman filters, whereas in reality the provided complete state estimate exclusively hangs on prior information—adding additional states and augmenting the filter does not help to update the information about the missing states.

- Only very rarely can optimal performance be achieved with strictly local controllers—unless one can predict the unobserved state.

- Adversary action can be correctly modeled using the game theoretic paradigm. The problem statement is then rigorous; however, there are few cases where solutions have been derived.

- Adversaries will resort to information warfare. The objective is to gain information about the true intention of the adversary without revealing any information. This is done by lying, deception, the use of decoys, and diversion tactics and gambits. This further complicates attempts at using game theoretic formulations to realistically address adversarial action.

## 2.8 Summary

The UAV cooperative team problem can be highly complex, even for relatively small teams. Generally the available theories and approaches can address only one or two aspects of the problem at a time. We are often more interested in a fast, feasible, and robust solution, rather than an optimal one. Since there are many UAV scenarios of moderate size, say, four to eight vehicles, that are of interest, approaches such as MILP and stochastic dynamic programming may be sufficiently fast at this scale, where a centralized optimal solution is possible. Thus, algorithm scalability may not be the limiting factor.

If more decentralization is desired, the primary limitation is task coupling. Task coupling can be reduced if a myopic or receding horizon procedure is used where not all tasks are addressed up front. However, this can have a significant impact on mission performance and even feasibility. Also, in the drive for localization, algorithms such as auction and distributed constraint satisfaction can incur extensive message traffic for all but the weakest task coupling. Finally, false information and communication delays can completely negate the benefits of cooperation—similar to losing the benefits of feedback when the sensors are noisy, and consequently open loop control is preferable.

## Bibliography


[1] M. Alighanbardi, Y. Kuwata, and J. How. Coordination and control of multiple UAVs with timing constraints and loitering. In *Proceedings of the* 2003 *American Control Conference*, Denver, CO, 2003.

[2] E. Balas and M. Carrera. A dynamic subgradient based branch and bound approach for set covering. *Operations Research*, 44:875–890, 1996.

[3] E. Balas and M. W. Padberg. Set partitioning: A survey. *SIAM Review*, 18:710–760, 1976.

[4] D. Bernstein, S. Zilberstein, and N. Immerman. The complexity of decentralized control of Markov decision processes. *Mathematics of Operations Research*, 27(4):819–840, 2003.

[5] D. Bertsekas. The auction algorithm: A distributed relaxation method for the assignment problem. *Annals of Operations Research*, 14:105–123, 1988.

[6] D. Bertsekas. Auction algorithms for network flow problems: A tutorial introduction. *Computational Optimization and Applications*, 1:7–66, 1992.

[7] D. Bertsekas. *Nonlinear Programming*. Athena Scientific, Nashua, NH, 1995.

[8] D. Bertsekas and D. Castañon. Parallel synchronous and asynchronous implementations of the auction algorithm. *Parallel Computing*, 17:707–732, 1991.

[9] D. Bertsekas and D. Castañon. Parallel primal-dual methods for the minimum cost network flow problem. *Computational Optimization and Applications*, 2:319–338, 1993.

[10] D. Bertsekas and D. Castañon. The auction algorithm for transportation problems. *Annals of Operations Research*, 20:67–96, 1989.

[11] D. Bertsekas and D. Castañon. Parallel asynchronous Hungarian methods for the assignment problem. *ORSA Journal on Computing*, 5(3):261–274, 1993.

[12] D. Bertsekas and J. Tsitsiklis. *Parallel and Distributed Computation: Numerical Methods*. Prentice–Hall, Englewood Cliffs, NJ, 1989.

[13] D. Bertsekas, D. Castañon, and H. Tsaknakis. Reverse auction and the solution of inequality constrained assignment problems. *SIAM Journal on Optimization*, 3:268–299, 1993.

[14] D. Bertsekas, D. Castañon, J. Eckstein, and S. Zenios. Parallel computing in network optimization. In *Network Models*. Handbooks in Operations Research 7, M. Ball, T. Magnanti, C. Monma, and G. Nemhauser, eds., 331–396, 1995.

[15] R. Burkard and R. Cela. *Linear Assignment Problem and Extensions*. Technical Report 127. Karl-Franzens University, Graz, Austria, 1998.

[16] C. Cassandras and W. Li. A receding horizon approach for solving some cooperative control problems. In *Proceedings of the 41st IEEE Conference on Decision and Control*, Las Vegas, NV, 2002.

[17] G. Dantzig. *Linear Programming and Extensions*. Princeton University Press, Princeton, NJ, 1963.

[18] C. Deissenberg and F. Gonzales. Pareto improving cheating in an economic policy game. *Computing in Economics and Finance*, Num 88, April 2001.

[19] L. J. Ford and D. Fulkerson. *Flows in Networks*. Princeton University Press, Princeton, NJ, 1962.

[20] P. Freeman. The secretary problem and its extensions: A review. *International Statistical Review*, 51:189–206, 1983.

[21] R. Gallager. *Information Theory and Reliable Communication*. Wiley, New York, 1968.

[22] A. R. Girard, M. Pachter, and P. Chandler. Decision making under uncertainty and human operator model for small UAV operations. In *Proceedings of the AIAA Guidance, Navigation, and Control Conference*, Keystone, CO, Aug. 2006.

[23] A. Goldberg and R. Tarjan. Solving minimum cost flow problems by successive approximation. *Mathematics of Operations Research*, 15:430–466, 1990.

[24] T. Groves. Incentives in teams. *Econometrica*, 41(4):617–631, 1973.

[25] W. Guo and K. Nygard. Combinatorial trading mechanism for task allocation. In *13th International Conference on Computer Applications in Industry and Engineering*, June 2001.

[26] Y. Ho and K. Chu. Team decision theory and information structures in optimal control problems, part 1. *IEEE Transactions on Automatic Control*, AC-17:15–22, 1972.

[27] A. Jadbabaie, J. Lin, and A. S. Morse. Coordination of groups of mobile autonomous agents using nearest neighbor rules. *IEEE Transactions on Automatic Control*, 48(6):988–1001, 2003.

[28] F. Jensen. *An Introduction to Bayesian Networks*. Springer-Verlag, Berlin, 1996.

[29] D. Kempka, J. Kennington, and H. Zaki. Performance characteristics of the Jacobi and the Gauss-Seidel versions of the auction algorithm on the Alliant FX/8. *ORSA Journal on Computing*, 3(2):92–106, 1991.

[30] H. Kuhn. *Extensive Games and the Problem of Information*, Annals of Mathematics Studies 28, Princeton University Press, Princeton, NJ, 1953.

[31] L. Lamport. The part-time parliament. *ACM Transactions on Computer Systems*, 16(2):133–169, 1998.

[32] W. Lovejoy. A survey of algorithmic methods for partially observable Markov decision processes. *Annals of Operations Research*, 28(1):47–66, 1991.

[33] R. Luce and H. Raiffa. *Games and Decisions: Introduction and Critical Survey*. Dover, New York, 1989.

[34] J. Marschak and R. Radner. *Economic Theory of Teams*. Yale University Press, New Haven, CT, 1972.

[35] D. Mayne and L. Michalska. Receding horizon control of nonlinear systems. *IEEE Transactions on Automatic Control*, AC-35(7):814–824, 1990.

[36] T. McLain and R. Beard. Coordination variables, coordination functions, and cooperative timing missions. *AIAA Journal of Guidance, Control, and Dynamics*, 28(1):150–161, 2005.

[37] P. Modi, W. Shen, M. Tambe, and M. Yokoo. An asynchronous complete method for general distributed constraint optimization. In *Proceedings of the Autonomous Agents and Multi-Agent Systems Workshop on Distributed Constraint Reasoning*, 2002.

[38] R. A. Murphy. An approximate algorithm for a weapon target assignment stochastic program. In *Approximation and Complexity in Numerical Optimization: Continuous and Discrete Problems*, P. M. Pardalos, ed., 406–421, Kluwer Academic Publishers, Dordrecht, The Netherlands, 2000.

[39] G. Nemhauser and B. Wolsey. *Integer and Combinatorial Optimization*. Wiley, New York, 1999.

[40] K. Nygard, P. Chandler, and M. Pachter. Dynamic network optimization models for air vehicle resource allocation. In *Proceedings of the American Control Conference*, Arlington, VA, June 2001.

[41] R. Olfati-Saber and R. Murray. Consensus protocols for networks of dynamic agents. In *Proceedings of the American Control Conference*, Denver, CO, 2003.

[42] C. Papadimitriou and K. Steiglitz. *Combinatorial Optimization: Algorithms and Complexity*. Prentice–Hall, Englewood Cliffs, NJ, 1982.

[43] E. Parzen. *Modern Probability Theory and Its Applications*. Wiley, New York, 1960.

[44] K. M. Passino. *Biomimicry for Optimization, Control, and Automation*. Springer, New York, 2005.

[45] R. W. Pew and A. S. Mavor Editors. *Modeling Human and Organizational Behavior*. National Academy Press, Washington, DC, 1998.

[46] M. Puterman. *Markov Decision Processes: Discrete Stochastic Dynamic Programming*. Wiley Series in Probability and Statistics, Wiley, New York, 2005.

[47] M. Puterman. *Markov Decision Processes*. Wiley Interscience, New York, 1994.

[48] S. Rasmussen, P. Chandler, and C. Schumacher. Investigation of single vs multiple task tour assignments for UAV cooperative control. In *Proceedings of the AIAA Guidance, Navigation, and Control Conference*, Monterey, CA, Aug. 2002.

[49] S. Rasmussen, P. Chandler, J. Mitchell, C. Schumacher, and A. Sparks. Optimal vs. heuristic assignment of cooperative autonomous unmanned air vehicles. In *Proceedings of the AIAA Guidance, Navigation, and Control Conference*, Austin, TX, 2003.

[50] S. J. Rasmussen and T. Shima. Tree search for assigning cooperating UAVs to multiple tasks. *International Journal of Robust and Nonlinear Control*, 18(2):135–153, 2008.

[51] G. Reinelt. *The Traveling Salesman Problem: Computational Solutions for TSP Applications*. Springer-Verlag, Berlin, 1994.

[52] A. Richards and J. How. Aircraft trajectory planning with collision avoidance using mixed integer linear programming. In *Proceedings of the American Control Conference*, Anchorage, AK, 2002.

[53] A. Richards, T. Bellingham, and J. How. Coordination and control of multiple UAVs. In *Proceedings of the AIAA Guidance, Navigation, and Control Conference*, Monterey, CA, Aug. 2002.

[54] S. Ross. *Introduction to Stochastic Dynamic Programming*. Academic Press, New York, 1983.

[55] T. Sandholm. An implementation of the contract net protocol based on marginal cost calculations. In *Proceedings of the 11th National Conference on Artificial Intelligence (AAAI-93)*, Washington, DC, 1993.

[56] C. Schumacher, M. Pachter, P. Chandler, and L. Pachter. UAV task assignment with timing constraints via mixed-integer linear programming. In *Proceedings of the AIAA 3rd Unmanned Unlimited Systems Conference*, 2004.

[57] C. Schumacher, P. Chandler, M. Pachter, and L. Pachter. Optimization of air vehicles operations using mixed integer linear programming. *Journal of the Operations Research Society*, 58(4):516–527, 2007.

[58] T. Shima and C. Schumacher. Assigning cooperating UAVs to simultaneous tasks on consecutive targets using genetic algorithms. *Journal of the Operational Research Society*, 2008.

[59] T. Shima, S. J. Rasmussen, A. Sparks, and K. Passino. Multiple task assignments for cooperating uninhabited aerial vehicles using genetic algorithms. *Computers and Operations Research*, 33(11):3252–3269, 2005.

[60] R. G. Smith and R. Davis. Frameworks for cooperation in distributed problem solving. *IEEE Transactions on Systems, Man, and Cybernetics*, 11(1):61–70, 1981.

[61] K. Sycara and J. Liu. Multiagent coordination in tightly coupled task scheduling. In *Proceedings of the International Conference on Multi-Agent Systems*, Kyoto, Japan, 1996.

[62] P. Toth and D. Vigo. *The Vehicle Routing Problem*. SIAM, Philadelphia, 2002.

[63] N. Vorobev. *Game Theory*. Springer-Verlag, New York, 1977.

[64] J. Wein and S. Zenios. Massively parallel auction algorithms for the assignment problem. In *Proceedings of the 3rd Symposium on the Frontiers of Massively Parallel Computation*, Nov. 1990.

[65] M. Wellman and P. Wurman. Market-aware agents for a multiagent world. *Robotics and Autonomous Systems*, 24:115–125, 1998.

[66] M. Yokoo and K. Hirayaman. Algorithms for distributed constraint satisfaction: A review. *Autonomous Agents and Multi-Agent Systems*, 3(2):198–212, 2000.

# Chapter 3

# Single-Task Tours

*Corey Schumacher and Tal Shima*

*Light is the task where many share the toil.*
—Homer

In this chapter we start dealing with the problem of assigning cooperating multiple UAVs to multiple targets. The scenario of interest involves wide-area search munitions (WASM), where multiple homogeneous UAVs must cooperatively perform multiple tasks on multiple targets. It is shown how the assignment problem can be solved using a capacitated transshipment problem formulation while assigning at each stage at most a single task to each vehicle.

## 3.1 Wide-Area Search Munition Scenario

The WASM scenario is a good starting point for the study of cooperative decision and control. This scenario exhibits many of the essential features of multiagent cooperation that are encountered in UAV teaming problems, without some of the additional complications inherent in combat UAV cooperation problems, such as threat avoidance.

### 3.1.1 Scenario Description

WASM are small powered UAVs, each with a turbojet engine and sufficient fuel to fly for a moderate duration (on the order of 30 minutes). A typical example is the Lockheed Martin vehicle shown in Fig. 3.1. The WASM are deployed in groups from larger aircraft flying at higher altitudes. Individually, the munitions are capable of searching for, recognizing, and attacking targets. When employed as a cooperative team instead of individuals, they can execute missions much more efficiently. Consequently, the ability to communicate information to one another is essential for maximum mission effectiveness. The vehicles

**Figure 3.1:** WASM. *(Courtesy of Lockheed Martin.)*

communicate known target information across the team whenever this information is updated by any vehicle, such as when a new target is found.

### 3.1.2 Required Tasks

We begin with a set of $N_v$ simultaneously deployed vehicles, each with a life span of 30 minutes. Let

$$V = \{1, 2, \ldots, N_v\} \tag{3.1}$$

be the set of deployed UAVs. Targets that might be found by searching fall into known classes according to the value or "score" associated with destroying them. We index them with $n$ as they are found, so that $n = 1, \ldots, N_t$, where $N_t$ is the number of discovered targets and

$$T = \{1, 2, \ldots, N_t\} \tag{3.2}$$

is the set of these targets. Let $V_n$ denote the value of target $n$. The WASM are assumed to begin a mission with no information about specific target numbers or locations. They do, however, have information about potential target types. The WASM are assumed to have a laser detection and ranging (LADAR) sensor that can scan a region in front of the vehicle, and automatic target recognition (ATR) software capable of detecting and identifying targets with some given probability from the resulting sensor information. The ATR process is modeled using a system that provides a probability or "confidence" that the target has been correctly classified. The probability of a successful classification is based on the viewing angle of the vehicle relative to the target [1]. Targets are not attacked unless a very high degree of confidence in target identification is achieved. In the simulation results presented here, a

90% probability of correct identification is required. A discussion of sensor modeling, false target attack rates, and how these issues affect cooperation can be found in Chapter 7.

### Task Definitions

The WASM are capable of performing four distinct tasks: search, classification, attack, and verification. Which tasks are required at any point in time is dependent on the available target information at that moment.

Search involves using the LADAR sensor to find unknown targets. Since cooperative search has been extensively studied [8], it will not be addressed here. For the purposes of our cooperative task assignment work, we assume an a priori specified search pattern, e.g., the lawnmower search, in which the vehicles have preassigned search lanes which, if completed, guarantee full coverage of the search area. If a WASM leaves its search path to perform other tasks, when it returns to search, it will return to the point where it left the search pattern, and then continue. Once a potential target, or *speculative*, has been located, a classification task is required. A classification is a second (or third) imaging and ATR of the target, from a different view angle, to raise the confidence level of the target identification. Once a speculative has been classified as an actual target, an attack task is needed. To perform an attack, the WASM dives low over the target and explodes, destroying itself and, hopefully, the target as well. The target is then imaged by another WASM to perform verification and confirm that the target has, in fact, been destroyed. If the target is still alive, another WASM may be tasked to attack it. Let

$$M = \{C, A, V\} \tag{3.3}$$

be the set of tasks on each target where $C$, $A$, and $V$ denote classify, attack, and verify, respectively.

### Precedence and Task Constraints

The three tasks needed to prosecute each target (classify, attack, and verification) must be performed in the specified order. Consequently, the cooperative planning algorithms must ensure that the proper order is achieved, with possible additional delays required between tasks, such as a delay between classify and attack for ATR, and a sufficient delay between attack and verification to allow any smoke or debris to clear enough to allow imaging. Additionally, each task the WASM may perform has unique constraints. Thus, it is not possible to simply assign three WASM to a target and allow each vehicle to perform whatever task is needed at the time it arrives. Classification requires specific look angles, dependent on the target orientation and previous view angle, and has a stand-off distance equal to the distance the sensor scans ahead of the WASM. Attack can be performed from any approach angle but requires the WASM to dive onto the target after locating it with the LADAR sensor. Verification is performed at the sensor stand-off distance, similar to classification, but can be performed from any approach angle.

A state transition diagram representation of the general problem is given in Fig. 3.2 where $p_c$, $p_k$, and $p_v$ denote the probability of performing a successful classification, attack, and verification, with 0.9, 0.8, and 0.95, respectively, being the thresholds chosen in this example.

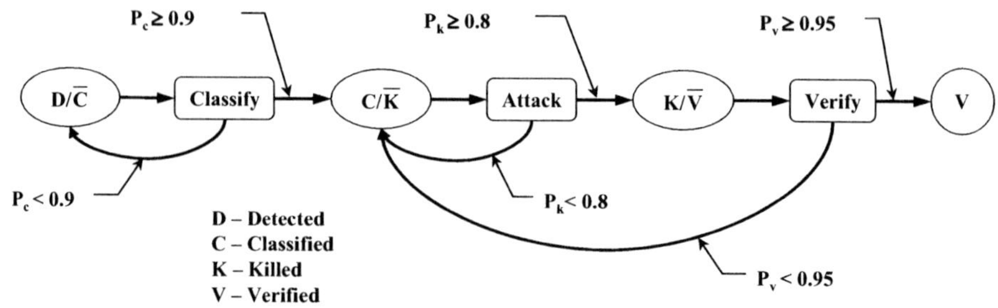


**Figure 3.2:** Target state transition diagram.

## 3.2 Capacitated Transshipment Assignment Problem Formulation

Whenever a new target is found, or an unassigned task is otherwise required (e.g., due to verification determining that a previously attacked target is not yet destroyed), an assignment algorithm is run to determine which members of the UAV team will perform the required tasks. For the simplest cooperative planning algorithm we will discuss, only a single task is planned for each known target. When that task has been performed, the next task is assigned. Eventually, all required tasks will be performed on a target, unless the target is of sufficiently low priority and the UAVs are instead assigned to continue searching for unknown targets, or the UAVs run out of fuel.

For this version of the assignment problem, we use a linear programming formulation known as a capacitated transshipment assignment problem (CTAP). The CTAP is used as a time-phased network optimization model designed to perform task allocation for a group of powered munitions each time it is run. The model is run simultaneously on all munitions at discrete points in time and assigns each vehicle up to one task each time it is run. The model is solved each time new information is brought into the system, typically because a new target has been discovered or an already-known target's status has been changed. This linear program can be described by a network flow diagram, as shown in Fig. 3.3. With $N_v$ UAVs and $N_t$ known targets, there are $N_v$ sources, $N_t$ potential nonsearch tasks, and up to $V$ possible search tasks. Each vehicle is assigned one task: search, classify, attack, or verification.

This simple assignment algorithm looks only at presently required tasks, with no knowledge of tasks likely to be needed in the future. For example, an attack task is not assigned until after a classify has been successfully performed. Similarly, a verification task is not assigned until after an attack is completed, even though we know that a verification will usually be desired after an attack. The advantage of this myopic approach is that the optimization problem at each stage of the mission can be solved very quickly, allowing for easy real-time implementation. The disadvantage, as will be discussed later, is that known information about future tasks (e.g., the requirement for verification after an attack task) is not being exploited. However, the simple CTAP approach results in an effective task

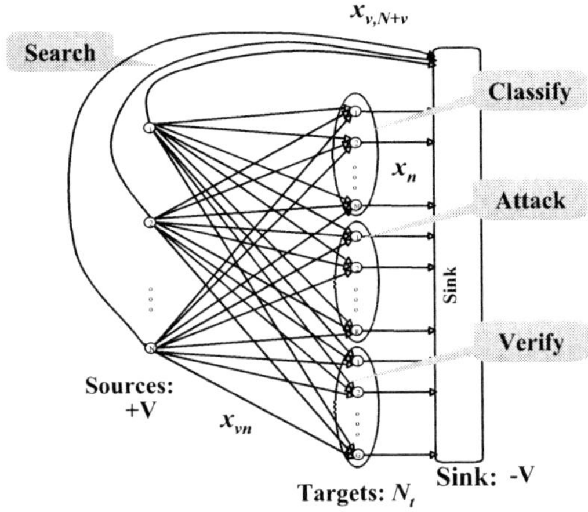


**Figure 3.3:** UAV CTAP network flow diagram.

assignment strategy for scenarios featuring task precedence constraints. This architecture naturally satisfies task precedence constraints by construction. Tasks are assigned only when they are "ready" to happen, so the precedence constraints are automatically met.

The CTAP network optimization model can be expressed as

$$\max J = \sum_{v,n} c_{vn}^k \, x_{vn} \tag{3.4}$$

subject to

$$x_{v,N+v} + \sum_{n=1}^{N} x_{vn} = 1 \qquad \forall v = 1, \ldots, N_v, \tag{3.5}$$

$$x_n - \sum_{v=1}^{V} x_{vn} = 0 \qquad \forall n = 1, \ldots, N_t, \tag{3.6}$$

$$\sum_{v} x_{v,v+n} + \sum_{n} x_n = N_t, \tag{3.7}$$

where $x \in \{0, 1\}$ is a binary decision variable; the variable $c_{vn}^k$ is the benefit of vehicle $v$ performing task $k$ on target $n$, where $k = 0$ corresponds to search, $k = 1$ corresponds to classify, $k = 2$ corresponds to attack, and $k = 3$ corresponds to verification. Search can only be paired with $n = 0$, the sink node. Referring back to Fig. 3.3, $x_{vn}$ refer to assignments of vehicles to individual targets, $x_{v,N+v}$ refer to the assignment of a vehicle to the search task, and $x_n$ is used to ensure flow balance constraints are met. Eq. (3.5) guarantees that each vehicle $v$ is assigned only one task. Eq. (3.6) is a flow balance constraint, requiring that for each flow into a task node $x_{vn}$, there is also one flow out, $x_n$. Eq. (3.7) requires that $N_t$ tasks be assigned. In combination with the requirement that each $x$ be binary, these constraints guarantee that each vehicle will be assigned only one task, either a search task or a target prosecution task, and also that each target prosecution task will be assigned to only one vehicle. It is possible for a target prosecution task to be left unassigned if it is of lower value than an available search task.

Note that in the CTAP formulation, there is only one potential source node for each vehicle, and thus there is no need for separate indices for node location and vehicle number. Likewise, there is only one task for each target, and thus there is no need for separate target and task indices. In the formulations in Chapters 4 and 5, separate indices for the vehicle and the task starting node will be necessary, and an additional index for the task to be performed will also be needed in the decision variables.

### 3.2.1 Weight Calculations

One of the critical questions involved in using the CTAP model for coordinated control and decision making for WASM is how the values of the benefits, or weights, $c_{vn}^k$ are chosen. Different values will achieve superior results for different situations. For example, reduced warhead effectiveness greatly increases the importance of verifying the success of an attack, and repeated attacks on an individual target that have not been destroyed. A simplified scheme has been developed which does not attempt to address the full probabilistic computation of the various expected values. It is intended to assign the highest value possible to killing a target of the highest-valued type, with other tasks generating less benefit. Overall, the chosen weights tend to result in the least possible lost search time for the execution of a particular task.

Next we discuss the primary issues involved in the selection of task weightings and present initial formulations for the various task weights. These equations will be further refined in later sections. The values of search, classify, attack, and verification tasks are all determined differently.

Suppose that there are $P$ distinct target types, each with a given value $V_p$ and probability of kill $P_{k_p}$ when attacked by a WASM, with $p = 1, \dots, P$. Let $\bar{V}$ be the maximum value of $V_p * P_{k_p} \, \forall p$ and let $\bar{p}$ be the value of $p$ corresponding to this maximum. Then $\bar{V}$ is the highest possible value for attacking a target.

A critical determinant in the resulting cooperative vehicle behavior is the ratio between the value of search and the value of target prosecution tasks. Since, for the WASM scenario, each vehicle is used up when it performs an attack task, the best that we can hope a team of $N$ WASM will do is attack $N$ targets of type $\bar{p}$. Therefore, the weights should be chosen so that the vehicles will attack any maximum-value targets identified, but they will not necessarily attack a lower-value target, especially when they have substantial fuel remaining. Accordingly, we set the value of search to start out equal to the value of attacking the highest value target type, and decrease continually as the vehicle fuel is used. Let $T_f$ be the remaining available flight time of the munition, and let $T_0$ be the initial endurance of the vehicle when launched. Then the value of search can be set to

$$c_{v0}^0 = \bar{V} * T_f / T_0. \tag{3.8}$$

This means that at any time after the initial launch, the value of search will always be lower than the value of attacking the highest valued target. Search value will decrease linearly with time, so that the decision algorithm will be more willing to conduct attacks and use up the available search agents.

Let $P_{id_n}$ be the confidence level with which the ATR process has identified target $n$. Then, the value of vehicle $v$ attacking target $n$ of type $p$ could be calculated as

$$c_{vn}^2 = P_{id_j} * P_{k_p} * V_p. \tag{3.9}$$

However, this choice for the attack weighting contains a flaw. Assuming heterogeneous vehicles, all vehicles will have the same value for conducting the attack task. A better formulation would give a preference to team members that can perform the task more quickly. Accordingly, we modify Eq. (3.9) as follows. Assume that vehicle $v$ will perform the attack on target $n$ at time $t_{vn}$. Then let

$$t_n^{min} = \min_v t_{vn} \tag{3.10}$$

for $i = 1, \ldots, n$. Then the modified value of vehicle $i$ attacking target $j$ can be expressed as

$$c_{vn}^2 = P_{id_n} * P_{k_p} * V_p * \frac{t_n^{min}}{t_{vn}}. \tag{3.11}$$

With this modification, the value for the attack with the shortest path is unchanged, while the value of other vehicles performing the task is lowered. Numerous variations are possible, such as raising the ratio of $t_n^{min}$ to $t_{vn}$ to an integer power to reduce the variation in values between vehicles, or, alternately, subtracting off the flight time required for each vehicle to reach the target and perform the attack. The time required for a vehicle to perform a specified task on a specified target is calculated using optimal path planning for a Dubins' car model and is discussed in Appendix B.

The value of vehicle $v$ attempting to classify target $n$ can be calculated as the assumed probability of classification, multiplied by the expected value of attacking the target, plus the value of the vehicle continuing to search after the classification task has been performed. Including the value of the remaining search time is critical for achieving the proper balance between search and classify task values. Thus, the value of vehicle $v$ attempting to classify target $n$, believed to be of type $p$, becomes

$$c_{vn}^1 = P_{cl} * P_{k_p} * V_p + \bar{V} * (T_f - T_{cl})/T_0, \tag{3.12}$$

where $T_{cl}$ is the amount of time that will be required for the vehicle to perform the classify task and then return to its search path and $P_{cl}$ is the expected probability of success for the classify task.

The value of a verification task is dependent on the likelihood that the target is still alive after an attack task. The value of vehicle $v$ performing verification on target $n$ of type $p$ can be expressed as

$$c_{vn}^3 = P_v * (1 - P_{k_p}) * P_{id_n} * V_p + \bar{V} * (T_f - T_{ver})/T_0, \tag{3.13}$$

where, similar to the classify case, $P_{ver}$ is the assumed probability of successful verification and $T_{ver}$ is the time required to perform the verification task and return to search. A more detailed version of the weightings can be generated if a confusion matrix is used, and multiple target types are possible with different probabilities. A more rigorous treatment of probabilistic impacts on cooperative team task assignment problems can be found in Chapter 7.

### 3.2.2 Simulation Results

This section presents simulation results for the CTAP algorithm applied to the WASM problem. In this example, eight vehicles are searching an area containing three targets of different types, and hence of different values. The target information is as follows:

| Target | Type | Value | Location (X,Y) |
|---|---|---|---|
| 1 | 1 | 10 | (4000,0) |
| 2 | 2 | 8 | (3500, −8500) |
| 3 | 1 | 10 | (18000, −11500) |

Target (X,Y) location is given in feet. The targets also have an orientation (facing) that has an impact on the ATR process and desired viewing angles, but this will not be discussed as it affects the task allocation only indirectly, through path length calculations. The search vehicles are initialized in row formation, with 15 minutes of flight time remaining, out of a maximum 30 minutes. This assumes that the vehicles have been searching for 15 minutes and then find a cluster of potential targets.

Determining precise appropriate values for the probabilities of successful ATR and verification requires substantial modeling of those processes and is beyond the scope of this work. Simplified models giving reasonable values for these parameters were used instead. The value of all possible tasks, vehicle, and target assignment combinations are calculated and sent to the capacitated transshipment problem solver. The values are multiplied by 10,000 before being sent to the solver, as it requires integers, and rounding will result in poor results without the scaling factor.

As vehicles are assigned nonsearch tasks, the possibility arises of missing targets, but that does not occur in this instance. We do not attempt to compensate for that possibility at this time. The simplest way to prevent missed targets would be to modify the preprogrammed search paths to close gaps created by vehicles that performed attack tasks. The vehicle and target locations shortly before the first targets are discovered are shown in Fig. 3.4(a). The open rectangles represent the sensor footprints of the searching vehicles, and the numbers are the vehicle locations. The lines trailing the vehicle numbers show flight paths. Targets are numbered 1 through 3.

The next snapshot, Fig. 3.4(b), is taken at $T = 34$ seconds. At this point, the searching munitions have discovered two of the targets, nearly simultaneously. When Target 2 was discovered, Vehicle 3 was assigned to upgrade the ATR classification on the target. After the classification on Target 2 was complete, Vehicle 3 was also assigned to attack it. Typically, the vehicle that performs the final ATR on a target will be the one assigned to destroy it, but this is not guaranteed. In particular, if one vehicle has significantly less fuel remaining than another, it is more likely to be assigned to destroy a target. Vehicle 8 is now assigned to image Target 3 and to verify that it has been destroyed. Vehicle 4 has been assigned to classify Target 1. In Fig. 3.4(b), two of the vehicle sensor footprints are shown in gray, to denote that the vehicle is banked and the sensor is thus inactive until the vehicle levels out again. All the other, unassigned vehicles are continuing to search.

In the next snapshot, taken at $T = 84$ seconds (Fig. 3.4(c)), the third target has been detected. Vehicle 4 has classified and attacked Target 1. Vehicle 8 has verified the destruction of Target 2 and is now proceeding to Target 1 to perform the same task. Vehicle 1 is imaging Target 3 to classify it and will then be assigned to attack it. Finally, Fig. 3.4(d)

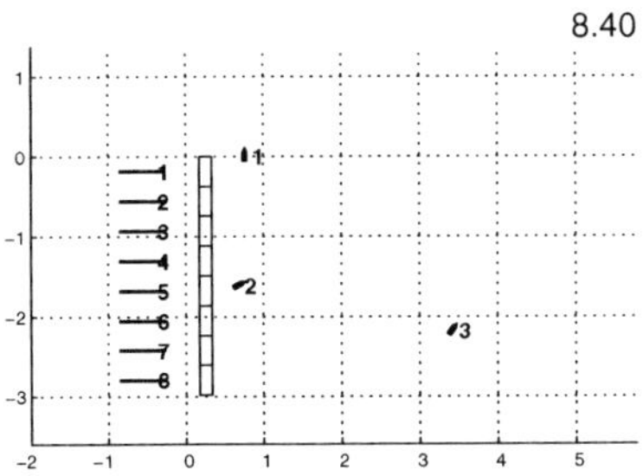


(a) Initial positions

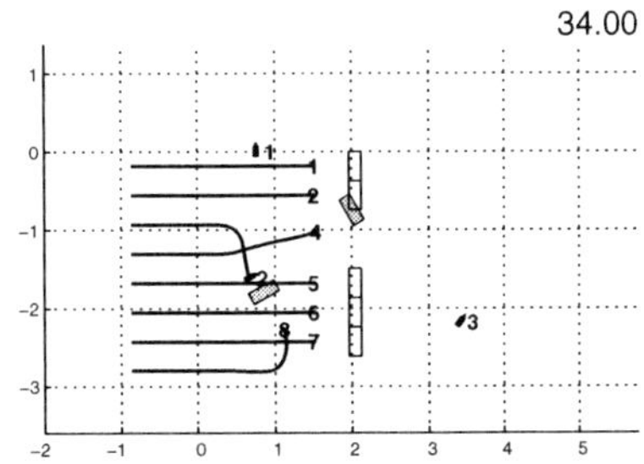


(b) First attack

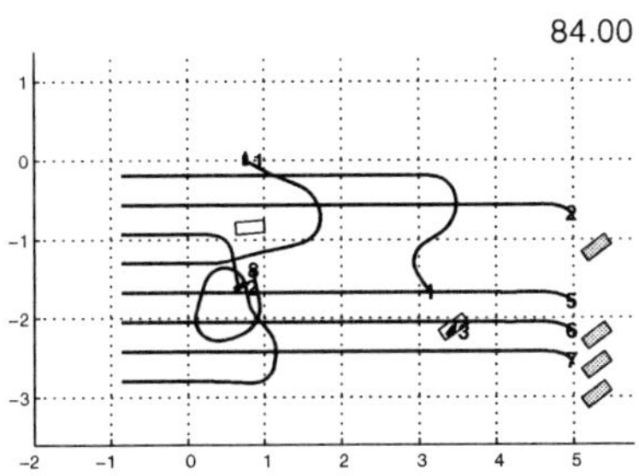


(c) Second attack

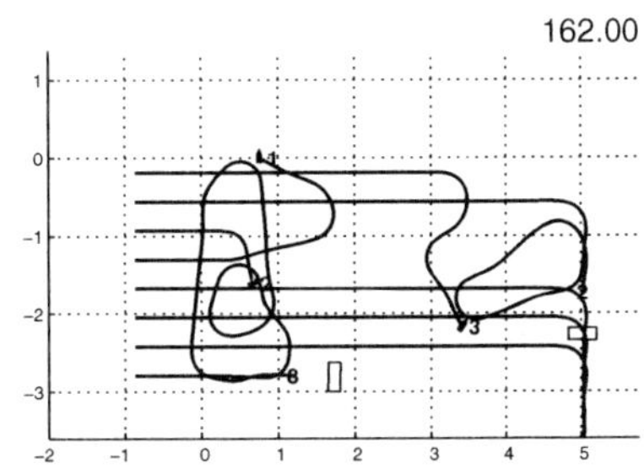


(d) Final attack

**Figure 3.4:** Vehicle paths for CTAP example.

shows the complete vehicle trajectories for prosecuting the three targets. Vehicle 2 has verified Target 3, and both Vehicles 2 and 8 have returned to their search patterns.

All discovered targets are fully prosecuted in this example. This will be the case for any discovered target, as long as it is of sufficient value. If the targets have a large amount of fuel remaining, resulting in a high value for the search task, then the vehicles will ignore low-value targets and search for higher-value targets, until their fuel reserve is low. An example of this behavior can be found in [7].

### 3.2.3 Capabilities and Limitations of CTAP

The CTAP formulation presented here for cooperative task assignment is a powerful and flexible tool. Many task assignment problems can be addressed with this method. The resulting optimization problems can be solved very rapidly, making it very amenable to real-time implementation [2]. For the problem sizes that could typically be encountered by eight vehicles, the assignment problem can be solved in a fraction of a second. The CTAP algorithm does, however, suffer from several significant limitations.

The first limitation is that all the necessary information needs to be gathered in one place, and at one time. The method cannot be implemented in an asynchronous or distributed fashion. The decision algorithm has to be run either in a centralized manner, where all computations are performed in one location and the resulting assignments are broadcast to all members of the team, or in an implicit coordination architecture, where the necessary information (team and target states, or task values) must be synchronized across all team members, so that they can each run an identical optimization problem.

A second limitation of CTAP is that it has limited flexibility. All the tasks that enter the assignment problem need to be independent. Only a very limited set of constraints can be included in the formulation. The CTAP is ideal for the classic $N$ on $N$ weapon-target assignment problem and for the WASM problem when search is considered as a possible task, again resulting in an $N$ on $N$ assignment problem. Some extensions are possible, such as the addition of possible tasks, resulting in an $N$ on $M$ assignment problem, with $M > N$, and only $N$ tasks assigned. Additionally, $N$ can be greater than $M$, with the problem then inverted and $M$ targets assigned to $N$ vehicles. Or additional tasks can be assigned to each target by the creation of dummy targets, as done in the previous section. However, the CTAP cannot directly incorporate any timing constraints, and it cannot directly account for coupling between tasks. The most obvious example of this is the task precedence constraint fundamental to the WASM scenario. Classification must be completed before attack, which must be completed before verification. There is no way to include these constraints directly into the CTAP. In fact, it is not possible to include the existence of the three tasks that are required for each target into the CTAP. Each target is responsible for only one task appearing in the CTAP at any stage, and the precedence constraints are guaranteed to be met by construction. However, the optimal decision algorithm contains no information about the staged nature of the tasks. This slows down task execution, as seen in Fig. 3.4(d), where Vehicle 2 is not assigned to verify Target 3 until after Target 3 has been attacked, resulting in slower completion of tasks than necessary. Much of the remainder of this book deals with the key question of how to incorporate additional coupling constraints into an optimal task assignment and scheduling process.

## 3.3 Iterative CTAP

A primary limitation of the CTAP assignment algorithm presented in section 3.2.3 is its inability to plan for the future. Known information about required future tasks cannot be incorporated into the algorithm. When any likely target is discovered, three tasks will probably be required: classification, attack, and verification. With the standard CTAP algorithm, the information about additional tasks is not used until all previous tasks have been completed. Only after classification is completed is attack considered, and only after an attack has taken place is verification considered.

The net result is a substantial delay in the execution of the verification task. This behavior is more easily seen in an animation of the vehicle paths but can be noted in the time-phased plots of Fig. 3.4. For example, Vehicle 2 performs verification on Target 3. However, this task is not assigned until after the attack is completed by Vehicle 1. As a result, Vehicle 4 does not leave its search path until the attack is complete and must fly back to Target 1. If this assignment had been made sooner, Vehicle 4 could have left its search path earlier and been in position to perform the verification immediately after the attack. This has the dual benefits of completing the prosecution of the target earlier, and reducing the time that the verification vehicle is away from its search pattern. Knowledge of likely future tasks can be incorporated into a CTAP assignment algorithm with a variety of iterative techniques. The key characteristic of such an iterative approach is that the CTAP algorithm is run repeatedly until all potential tasks for all known targets have been assigned or rejected as being less valuable than continued search. When a new target is discovered, the CTAP algorithm is run, including the next required task on all known targets. When that assignment has been calculated, some subset of the resulting assignment is kept. Vehicle and target states are updated assuming those assignments are carried out, and the assignment algorithm is run again. This iterative process is continued until all potential tasks are assigned or rejected in favor of search. Typically, this results in the CTAP algorithm being run at most one time for each available task. Since the CTAP can be solved very quickly, this methodology is still suitable for real-time implementation. Another option would be to save all the assignments from each CTAP run, which would usually assign all tasks within three cycles. The advantage of the first option is that it is likely to use less vehicle flight time up on off-search tasks, as one vehicle could be assigned many tasks, if that is most efficient. An advantage of the latter strategy is that all tasks are likely to be completed more quickly, as no vehicle will be assigned more than a few tasks, enabling quicker prosecution of the targets.

The iterative CTAP strategy typically results in improved performance if potential future tasks can be forecast with high probability. If the probability of a classification task resulting in an attack and a verification being performed is very high, then it is advantageous to plan ahead and assign two or three vehicles to begin those tasks. However, if the false target detection rate is very high, then it is disadvantageous to pull two or three vehicles off of search for every speculative target, as much search time would be wasted on false positives. In that case, it is better only to use one vehicle to confirm the target identity before calling in additional resources. The key point, which applies to other cooperative control scenarios and methods, is that the value of cooperation, and the value of planning ahead, are both fundamentally dependent on the quality of information available.

### 3.3.1 Simulation Example

A simulation example using the iterative CTAP algorithm, with a single vehicle-task assignment pairing kept from each iteration, is shown in Fig. 3.5. In this case, only one assignment is selected from each iteration. The assignment that would be completed soonest is selected, the virtual vehicle and target states are updated, and the CTAP algorithm is run again.

The targets are as follows:

| Target | Type | Value | Location (X,Y) |
|---|---|---|---|
| 1 | 1 | 10 | (5000,0) |
| 2 | 2 | 8 | (3500, −8500) |
| 3 | 1 | 10 | (18000, −11500) |

Target 1 has been moved east 1000 feet from the example in Fig. 3.4 to clearly differentiate the two assignment calculations performed as each target is found. Vehicle

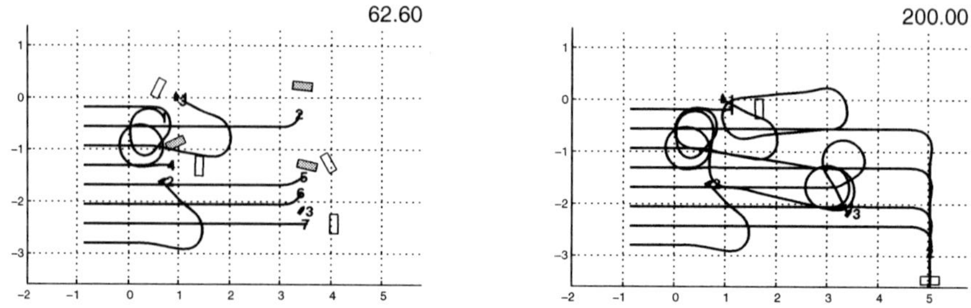


(a) $t = 62.6$ sec

(b) $t = 200$ sec

**Figure 3.5:** Vehicle paths for iterative CTAP example.

trajectories through the first 62.6 seconds are shown in Fig. 3.5(a), and the complete vehicle trajectories through 200 seconds are shown in Fig. 3.5(b). In this case, Target 2 is located first. Initially, Vehicle 4 is assigned to verify the attack on Target 2, but when Target 1 is found, and the assignment is run again, Vehicle 4 loses this task to Vehicle 1, which performs the verification on Target 2. Similarly, Vehicle 6 is assigned to verify Target 3 but then loses this assignment when Vehicle 1 is assigned to verify Target 3 instead. In both cases, this undesirable "churning" effect wastes vehicle time. All tasks on all discovered targets are still completed, but not in an efficient manner. The actual tasks were completed as follows: Vehicles 8 and 3 perform classify and attack on Targets 2 and 1, respectively. Vehicle 5 performs classify and attack on Target 3. Vehicle 2 performs verify on Target 1. Vehicle 1 performs verify on Targets 2 and 3.

### 3.3.2 Memory Weighting

As illustrated in the previous example, when a team of WASM encounter a new target while in the midst of executing a previously determined task plan, a replan is required. This replan may result in vehicles changing assigned tasks. When there is a significant benefit to doing so, this is a desirable result. However, the algorithm can be very sensitive to minor variations in task weights. One potential source of such variations can be replanning of the vehicle paths. The path planning is performed using the Dubins' car model for vehicle dynamics, which assumes the ability to instantly generate a desired turn rate. However, in the MultiUAV2 simulation (see Appendix A), we use a more accurate six-degree-of-freedom (6DOF) model to simulate the vehicles. As a result, they cannot fly the exact paths calculated for the Dubins' car model (see Appendix B). Therefore, when replanning, the vehicles typically will not be on precisely the planned trajectory. Due to discontinuities in the path planning process, these small variations in vehicle position can result in selection of a substantially different path and a discontinuous change in the value of a task.

Even when a new task is of higher value than a previously assigned task, it may not be in the best interest of the vehicle team for a member to switch off a task that is already in the process of being performed. The CTAP optimization being used does not contain

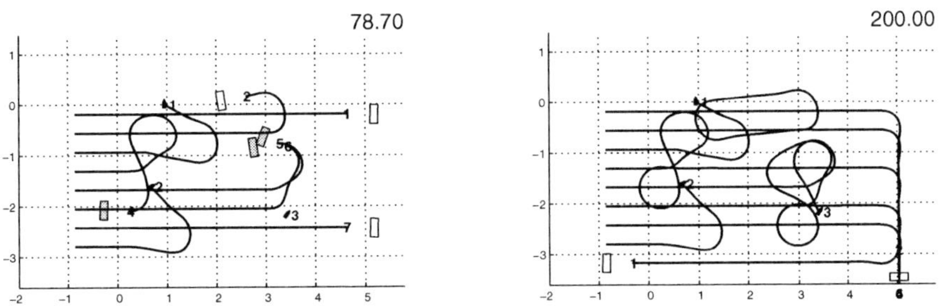


(a) $t = 62.6$ sec (b) $t = 200$ sec

**Figure 3.6:** Vehicle paths for iterative CTAP with memory.

any information about what tasks are already being performed or what has been assigned previously. The performance of the overall system can be improved by the addition of a memory weighting factor $c_{mem}$ to the task weights (Eqs. 3.11–3.13), resulting in

$$
\begin{aligned}
c_{vn}^1 &= (P_{cl} * P_{k_p} * V_p + \bar{V} * (T_f - T_{cl})/T_0) * c_{mem},\\
c_{vn}^2 &= (P_{id_j} * P_{k_p} * V_p * \tfrac{t_n^{min}}{t_{vn}}) * c_{mem},\\
c_{vn}^3 &= (P_{ver} * (1 - P_{k_p}) * P_{id_n} * V_p + \bar{V} * (T_f - T_{ver})/T_0) * c_{mem}.
\end{aligned}
$$

The additional weighting is used to encourage vehicles to continue on to service targets to which they have already been assigned, and thus reduce the churning effect which can occur if vehicle-target assignments change frequently. Simulation studies [4] have found that $c_{mem} = 1.05$ greatly reduces the undesirable churning while still allowing the needed flexibility to switch to newly available high-value tasks. It is important to note that $c_{mem} = 1.05$ only for the vehicle-task pairings for which the vehicle is already in the process of performing the task, and $c_{mem} = 1.00$ for all other vehicle-task pairings.

The simulation example shown in Fig. 3.5 can be run again with this memory weighting added to the task weights. The results are shown in Fig. 3.6, again with one plot partway through the run and a second plot showing the complete vehicle trajectories. With the small additional vehicle memory weight on the tasks a vehicle is currently in the process of performing, neither Vehicle 4 nor Vehicle 6 loses its initial verification task assignment, resulting in less wasted vehicle flight time due to churning and, overall, more efficient task execution.

### 3.3.3 Path Elongation

An additional complication arises from the use of the Iterative CTAP strategy. In the original CTAP formulation, the minimum length path is calculated for each potential vehicle-task pairing. Since the previous task has already been completed, it is impossible for the vehicle assigned to perform the next task to arrive early. However, with iterative CTAP, all three potential tasks needed to prosecute a target can be assigned immediately. It is essential to ensure that the task precedence and timing constraints are met. One method is simply to use the minimum length paths, check the completion time for each task, and assign a zero value to any vehicle-task pairing that would occur *before* any required tasks. Thus, if a calculated verification path would arrive before the assigned attack is completed, that vehicle-task pairing is assigned a zero value and would not be assigned. This approach guarantees that precedence constraints will be satisfied but does not guarantee that all tasks will be assigned. A better strategy is to calculate extended flight paths that will meet the precedence and timing constraints. Algorithms for calculating paths meeting specified position, heading, and time constraints for Dubins' car model have been developed and can be used to calculate the necessary paths [5, 6]. In this case, whenever a minimum-length path would arrive too early, that path is extended so that the task is completed just after the preceding task. This guarantees that all tasks will be assigned if they are of sufficiently higher value than the search task.

## 3.4 Summary

In this chapter we have started with the study of cooperative decision and control algorithms for the WASM scenario. The problem was posed and it was shown and discussed how CTAP can be applied for the solution. Next it was shown that the iterative CTAP with memory weighting is a powerful tool for solving such multiple vehicle assignment problems as it enables solving the problem very rapidly and is appropriate for real-time implementation. Monte Carlo results have indicated that compared to CTAP, the iterative CTAP results, on average, in 20% improved costs [3]. However, the greedy, iterative nature of the algorithm prevents any performance guarantees in an individual case. In the next section, we will examine strategies for the optimal assignment and scheduling of cooperative WASM tasks, while taking into account from the outset the required multiple tasks on each target.

## Bibliography

[1] P. Chandler and M. Pachter. Hierarchical control of autonomous teams. In *Proceedings of the AIAA Guidance, Navigation, and Control Conference*, Montreal, Canada, 2001.

[2] K. E. Nygard, P. R. Chandler, and M. Pachter. Dynamic network flow optimization models for air vehicle resource allocation. In *Proceedings of the American Control Conference*, Arlington, VA, 2001.

[3] S. J. Rasmussen, P. R. Chandler, J. W. Mitchell, C. J. Schumacher, and A. G. Sparks. Optimal vs. heuristic assignment of cooperative autonomous unmanned air vehicles. In *Proceedings of the AIAA Guidance, Navigation, and Control Conference*, Austin, TX, 2003.

[4] C. Schumacher, P. Chandler, and S. Rasmussen. Task allocation for wide area search munitions via iterative network flow. In *Proceedings of the AIAA Guidance, Navigation, and Control Conference*, Monterey, CA, 2002.

[5] C. Schumacher, P. Chandler, S. Rasmussen, and D. Walker. Task allocation for wide area search munitions with variable path length. In *Proceedings of the American Control Conference*, Denver, CO, 2003.

[6] C. Schumacher, P. Chandler, S. Rasmussen, and D. Walker. Path elongation for UAV task assignment. In *Proceedings of the AIAA Guidance, Navigation, and Control Conference*, Austin, TX, 2003.

[7] C. J. Schumacher, P. R. Chandler, and S. J. Rasmussen. Task allocation for wide area search munitions via network flow optimization. In *Proceedings of the AIAA Guidance, Navigation, and Control Conference*, Montreal, Canada, 2001.

[8] L. D. Stone. *Theory of Optimal Search*, 2nd ed. Military Applications Society, Arlington, VA, 2004.

# Chapter 4
# Multiple Assignments

*Tal Shima, Corey Schumacher, and Steven Rasmussen*

*Let everyone sweep in front of his own door, and the whole world will be clean.*
—Johann Wolfgang von Goethe

Assigning vehicles to targets with single task tours makes it necessary to solve some multiple-vehicle multiple-target assignment problems in a greedy manner by using strategies such as recalculating the assignments for all vehicles after each vehicle finishes its currently assigned task. This is inefficient and can introduce delays in the system that lead to phenomena such as churning; see Chapter 3, section 3.3.1. Since the required tasks are known a priori, plans can be generated to assign multiple tasks to each vehicle, i.e., multiple task tours. In this chapter we discuss methods of assigning multiple task tours to vehicles. These methods include mixed integer linear programming (MILP), tree search, and genetic algorithms (GAs).

## 4.1 Mixed Integer Linear Programming

The fundamental limitation of the CTAP algorithms presented previously is that all the decision variables are binary. Solutions can be computed very quickly, but only a very limited set of problems can be directly formulated in such a manner. A natural extension is to the MILP, where both integer and real-valued decision variables can be included. The use of MILP allows rigorous inclusion of timing constraints and opens up the possibility of optimal solutions for a much wider range of joint task assignment and scheduling problems.

The WASM cooperative control problem will now be solved with a MILP model using a discrete representation of the real world based on nodes that represent discrete start and end positions for segments of a vehicle's path. A similar formulation, with only minor

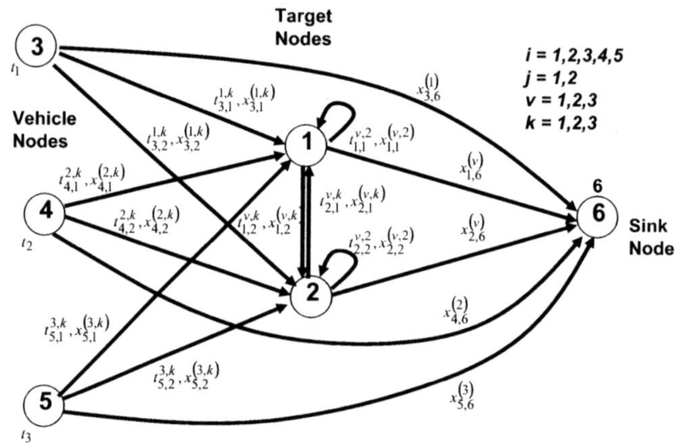


**Figure 4.1:** State transition diagram for two targets, three vehicles.

variations, could be applied to other cooperative UAV problems, and we will thus refer to generic UAVs instead of WASM henceforth.

Consider $N$ geographically dispersed targets with known positions and $V$ UAVs. We assume $V \geq N + 1$. We then have $N + V + 1$ nodes: $N$ target nodes, $V$ source nodes (the points of origin of the UAVs), and one sink node. Nodes $1, \ldots, n$ are located at the $N$ target positions. Nodes $N+1, \ldots, N+V$ are located at the initial positions of the vehicles. Node $N + V + 1$ is the "sink." The sink node is used when no further assignment for the UAV is in the offing; it goes to the sink when it is done with all of its tasks, or when it is not assigned another task. In practice, when a UAV enters the sink it is then used for performing search of the battle space. The MILP model requires the information on the costs (or times) for a UAV to fly from one node to another node. These known flight times are constraints represented by $t_{in}^{vk} \geq 0$, the time it takes UAV $v$ to fly from node $i$ to node $n$ to perform task $k$. A node diagram for the possible assignments is shown in Fig. 4.1.

The indices are $i = 1, \ldots, N + V$, $n = 1, \ldots, N$, $v = 1, \ldots, V$, and $k = 1, \ldots, K$. The index $k$ designates the task to be performed at node $n$. Thus, it is acknowledged that the time to travel from node $i$ to node $n$ depends on the particular UAV's airspeed and the assigned task $k$. For the WASM problem, $K = 3$ tasks must be performed on each target:

1. $k = 1$, classification,
2. $k = 2$, attack,
3. $k = 3$, verification.

Furthermore, once a UAV attacks a target, it is destroyed and can no longer perform additional tasks. This is certainly the case for powered munitions and the WASM mission, but if the UAV was a reusable aircraft, one would have to modify the problem formulation and account for the UAVs' depletion of its store of ammunition following each attack.

The three tasks must be performed on each target in the order listed. This results in critical timing constraints, which set this problem apart from the classical vehicle routing problem (VRP) [16]. In the latter, rigid time windows for the arrival of the vehicles can be

**Table 4.1:** MILP decision variables.

| Variable | Type | No. Variables | Meaning |
|---|---|---|---|
| $x_{in}^{vk}$ | binary | $NV(3N+1)$ | Assignment of vehicle $v$ at node $i$ to perform task $k$ on target $n$ |
| $x_{i,N+V+1}^{v}$ | binary | $(N+1)V$ | Assignment of vehicle $v$ at node $i$ to search |
| $t_n^k$ | continuous | $3N$ | Time task $k$ is completed on target $n$ |
| $t_v$ | continuous | $V$ | Duration vehicle $v$ loiters before starting its task tour |

specified; however, the coupling brought about by the need to sequence the various tasks is absent. Evidently, our problem features some aspects of job shop scheduling [7]. Finally, the endurance of UAV $v$ is $T_v$.

### Decision Variables

The binary decision variable $x_{in}^{vk} = 1$ if UAV $v$ is assigned to fly from node $i$ to node $n$ and perform task $k$ at node $n$, and 0 otherwise, with $i = 1, \dots, N+V$, $n = 1, \dots, N$, $v = 1, \dots, V$, and $k = 1, \dots, K$. For $i = N+1, \dots, N+V$, only $x_{N+v,n}^{vk}$ exist. These variables correspond with the first task assignment each vehicle receives, starting from its unique source node. Only vehicle $v$ can do a task starting from node $N+v$. For task assignments $k = 1, 3$, $i \neq n$, and for task assignment $k = 2$ we allow $i = n$. The latter allows for a UAV to perform the target classification task and immediately thereafter attack the target. Thus far, we have $NV(3N+1)$ binary decision variables. We also have additional binary decision variables corresponding to the search task. The decision variable $x_{i,N+V+1}^{v} = 1$ if UAV $v$ is assigned to fly from node $i$ to the sink node $N+V+1$ and is 0 otherwise, with $v = 1, \dots, V$ and $i = 1, \dots, N+V$. This adds $(N+1)V$ binary decision variables. Being assigned to the sink node can also be thought of as being assigned to the search task.

Continuous decision variables are used to represent the times that actions will occur. The time of performance of task $k$ on target $n$ is $t_n^k$; $k = 1, 2, 3$ and $n = 1, \dots, N$. Thus, we have $3N$ continuous decision variables. We also have $V$ additional continuous decision variables: the time UAV $v$ leaves node $n = N+v$ is $t_v$; $v = 1, \dots, V$. In total we have $V[N(3N+2)+1]$ binary decision variables and $3N+V$ continuous nonnegative decision variables. The binary and continuous decision variables used in the WASM MILP formulation are summarized in Table 4.1.

## 4.1.1 Cost Functions

A variety of cost functions are possible, and the chosen cost function must be determined based on the details of the desired UAV team response. A few useful cost functions are the following:

1. Minimize the total flight time of the UAVs:

$$J = \sum_{k=1}^{K} \sum_{v=1}^{V} \sum_{i=1}^{N+V} \sum_{n=1}^{N} t_{i,n}^{v,k} x_{i,n}^{v,k}, \tag{4.1}$$

where $t_{in}^{vk}$ is the time required for vehicle $v$ to travel from its present node, $i$, to the target node $n$ to perform task $k$.

2. Alternately, minimize the total engagement time. Target $n$ is visited for the last time at time $t_n^3$. Let $t_f$ be the time at which all targets have been verified. Introduce an additional continuous decision variable $t_f \in \Re_+^1$. The cost function is then $J = t_f$ and we minimize $J$ subject to the additional constraint:

$$t_n^3 \le t_f, \qquad n = 1, \dots, N. \tag{4.2}$$

This cost function minimizes the total time taken to prosecute all the targets. However, it does not provide any penalty for delays on the prosecution of individual targets, as long as the prosecution time for the last target is not delayed. Accordingly, improved performance can be gained by adding a small weight to encourage the early prosecution of each individual target. Then the cost function becomes

$$J = t_f + \sum_{n=1}^{N} c_n^3 t_n^3, \tag{4.3}$$

where $c_n^k$ is a weight on the completion time of task $k$ on target $n$. Thus, $c_n^3$ is a small positive weight on the verification time for each target $n$. This addition to the cost function encourages the completion of each verification task as early as possible, instead of only the time for the last task completion that affected the cost function.

### 4.1.2 Constraints

The most challenging aspect of representing the WASM task assignment problem by a MILP is in the formulation of appropriate constraints. Inclusion of all the necessary constraints is essential to enforce the desired vehicle behavior. A complete set of constraints includes both nontiming and timing constraints. These include the following:

1. All three tasks for each target must be completed exactly one time:

$$\sum_{v=1}^{V} \sum_{i=1, i \ne n}^{N+V} x_{in}^{vk} = 1, \quad n = 1, \dots, N, \quad k = 1, 3, \tag{4.4}$$

and

$$\sum_{v=1}^{V} \sum_{i=1}^{N+V} x_{in}^{v2} = 1, \quad n = 1, \dots, N. \tag{4.5}$$

2. Each UAV can visit the target node $N$, coming from the outside, at most one time:

$$\sum_{k=1}^{3} \sum_{i=1,i\neq n}^{N+V} x_{in}^{vk} \leq 1, \quad v = 1, \ldots, V, \quad n = 1, \ldots, N. \tag{4.6}$$

3. Additionally, each UAV $v$ can enter the sink only once:

$$\sum_{i=1}^{N+V} x_{i,N+V+1}^{v} \leq 1, \quad v = 1, \ldots, V. \tag{4.7}$$

4. Each UAV $v$ leaves node $n$ at most once:

$$\sum_{k=1}^{3} \sum_{i=1,i\neq n}^{N} x_{in}^{vk} \leq 1, \quad v = 1, \ldots, V, \quad n = 1, \ldots, N. \tag{4.8}$$

5. For the WASM problem, each UAV is used up when it performs an attack and therefore can attack at most one target:

$$\sum_{n=1}^{N} \sum_{i=1}^{N+V} x_{in}^{v2} \leq 1, \quad v = 1, \ldots, V. \tag{4.9}$$

6. If UAV $v$ is assigned to verify target $n$, it cannot be assigned to attack target $n$:

$$\sum_{i=1,i\neq n}^{N+V} x_{in}^{v2} \leq 1 - \sum_{i=1,i\neq j}^{N+V} x_{in}^{v3}, \quad v = 1, \ldots, V, \quad n = 1, \ldots, N. \tag{4.10}$$

The next several constraints enforce the appropriate flow balance at each node. Any vehicle that enters a target node must also exit it, unless it performed an attack task, which must be terminal.

7. If UAV $v$ enters target (node) $n$ to perform task 3, it must also exit target node $n$:

$$\sum_{i=1,i\neq n}^{N+V} x_{in}^{v3} \leq \sum_{k=1}^{3} \sum_{i=1,i\neq n}^{N} x_{ni}^{vk} + x_{n,N+V+1}^{v}, \quad v = 1, \ldots, V, \quad n = 1, \ldots, N. \tag{4.11}$$

8. If UAV $v$ enters target (node) $j$ to perform task 1, it must exit or attack target node $j$:

$$\sum_{i=1,i\neq n}^{N+V} x_{in}^{v1} \leq \sum_{k=1}^{3} \sum_{i=1,i\neq n}^{N} x_{ni}^{vk} + x_{nn}^{v} + x_{n,N+V+1}^{v}, \quad v = 1, \ldots, V, \; n = 1, \ldots, N. \tag{4.12}$$

9. A munition is perishable. If UAV $v$ is assigned to fly to target (node) $n$ to perform task $k = 2$, then, at any other point in time, UAV $v$ cannot also be assigned to fly from target $n$ to any other node. Recall that UAV $v$ can enter target $n$ not more than once. Thus it cannot both attack a target node $n$ and leave that node:

$$\sum_{k=1}^{3} \sum_{i=1, i \neq n}^{N} x_{ni}^{vk} + x_{n,N+V+1}^{v} \leq 1 - \sum_{i=1}^{N+V} x_{in}^{v2}, \quad v = 1, \ldots, V, \quad n = 1, \ldots, N. \tag{4.13}$$

10. A UAV $v$ cannot leave a node $n$ that it is not assigned to enter:

$$\sum_{k=1}^{3} \sum_{i=1, i \neq n}^{N} x_{ni}^{vk} + x_{n,N+V+1}^{v} \leq \sum_{k=1}^{3} \sum_{i=1, i \neq n}^{N+V} x_{in}^{vk}, \quad v = 1, \ldots, V, \ n = 1, \ldots, N. \tag{4.14}$$

11. All UAVs leave their source nodes, even if this entails a direct assignment to the sink.

$$\sum_{k=1}^{3} \sum_{n=1}^{N} x_{N+v,}^{vk} + x_{N+v,N+V+1}^{v} = 1, \quad v = 1, \ldots, V. \tag{4.15}$$

Assignment to the sink corresponds with default behavior, in this case, search.

12. A UAV cannot attack target $i$, coming from node $i$, unless it entered node $i$ to perform a classification:

$$x_{ii}^{v2} \leq \sum_{n=1}^{N+V} x_{ni}^{v1}, \quad i = 1, \ldots, N, \quad v = 1, \ldots, V. \tag{4.16}$$

This results in $3(N+V)+8NV$ nontiming constraints. Additional timing constraints are required.

13. Nonlinear equations which enforce the timing constraints are easily derived:

$$t_n^k = \sum_{v=1}^{V} \sum_{i=1, i \neq n}^{N+V} \left[ \left(t_i^1 + t_{in}^{vk}\right) \sum_{l=1, l \neq i}^{N+V} x_{li}^{v1} + \left(t_i^3 + t_{in}^{vk}\right) \sum_{l=1, l \neq i}^{N+V} x_{li}^{v3} \right] x_{in}^{vk}, \tag{4.17}$$

$$t_n^k = \sum_{v=1}^{V} \sum_{i=1}^{N+V} \left[ \left(t_i^1 + t_{in}^{v2}\right) \sum_{l=1, l \neq i}^{N+V} x_{li}^{v1} + \left(t_i^3 + t_{ij}^{v2}\right) \sum_{l=1, l \neq i}^{N+V} x_{li}^{v3} \right] x_{in}^{v2} \tag{4.18}$$

$\forall n = 1, \ldots, N;\ k = 1, 3.$

However, an alternative linear formulation is needed, so that a MILP formulation is achieved. A linear formulation for the timing constraints can be created using sets of linear inequality constraints. Thus, let

$$T \equiv \max_{v} \{T_v\}_{v=1}^{w} \tag{4.19}$$

be the maximum remaining flight time of any of the vehicles. Then the linear timing constraints can be expressed as

$$t_n^k \leq t_i^1 + t_{in}^{vk} + \left(2 - x_{in}^{vk} - \sum_{l=1, l \neq i}^{N+V} x_{l,i}^{v,1}\right) VT, \tag{4.20}$$

$$t_n^k \geq t_i^1 + t_{in}^{vk} - \left(2 - x_{in}^{vk} - \sum_{l=1, l \neq i}^{N+V} x_{l,i}^{v,1}\right) VT, \tag{4.21}$$

$$t_n^k \leq t_i^3 + t_{in}^{vk} + \left(2 - x_{in}^{vk} - \sum_{l=1, l \neq i}^{N+V} x_{l,i}^{v,3}\right) VT, \tag{4.22}$$

$$t_n^k \geq t_i^3 + t_{in}^{vk} - \left(2 - x_{in}^{vk} - \sum_{l=1, l \neq i}^{N+V} x_{l,i}^{v,3}\right) VT \tag{4.23}$$

for $i = 1, \ldots, N$, $n = 1, \ldots, N$, $i \neq n$, $v = 1, \ldots, V$, and $k = 1, 2, 3$.

This results in $6NV(N-1)$ timing constraints.

14. Additional linear timing constraints are needed for the case where $i = j$, i.e., when a vehicle $v$ both classifies and then attacks a target $n$:

$$t_i^2 \leq t_i^1 + t_{ii}^{v2} + \left(2 - x_{ii}^{v2} - \sum_{l=1, l \neq i}^{N+V} x_{l,i}^{v,1}\right) wT, \tag{4.25}$$

$$t_i^2 \geq t_i^1 + t_{ii}^{v2} - \left(2 - x_{ii}^{v2} - \sum_{l=1, l \neq i}^{N+V} x_{l,i}^{v,1}\right) wT \tag{4.26}$$

for $i = 1, \ldots, n$, $v = 1, \ldots, V$.

This gives additional $2NV$ timing constraints.

15. All of the above timing constraints apply to the case of a vehicle flying from one target node to another. An additional $6NV$ constraints are needed for the first task each vehicle performs:

$$t_n^k \leq t_v + t_{N+v,n}^{vk} + (1 - x_{N+v,n}^{vk}) VT, \tag{4.28}$$

$$t_n^k \geq t_v + t_{N+v,n}^{vk} - (2 - x_{N+v,n}^{vk}) VT \tag{4.29}$$

for $j = 1, \ldots, n$, $v = 1, \ldots, w$, and $k = 1, 2, 3$.

These timing constraints operate in pairs. They form loose, effectively inactive inequalities for assignments which are not made, but each pair results in a hard equality constraint for any assignment $x_{in}^{vk} = 1$ which occurs.

16. Lastly, task precedence constraints must be enforced:

$$t_n^1 \leq t_n^2, \tag{4.30}$$

$$t_n^2 \leq t_n^3 \tag{4.31}$$

$$\text{for } n = 1, \dots, N$$

resulting in $2N$ additional constraints.

Thus, for an $N$-target, $W$-vehicle problem, there are $3(N+V)+8NV$ nontiming constraints and $6N^2V + 2N(V+1)$ timing constraints.

## 4.2 Tree Search

An alternative approach to MILP is to represent the WASM scenario in the form of a decision tree. Graphs and trees are described in many papers and textbooks [3, 4] and will be only briefly reviewed in the next subsection. Then a search algorithm, proposed in [8], will be described.

### 4.2.1 Background on Decision Trees

A graph is defined as a set of *vertices* $B$ and a set of ordered pairings of those vertices $E$. Thus, the graph is defined as $G = (B, E)$. Vertices are also termed *nodes*, and each ordered pair of vertices in $E$ is called an *edge*. Edges can also be thought of as the connection between two vertices. A *directed* graph is a graph where the edges can be traversed only in a defined direction, i.e., the ordering of the vertices in the edges determines the direction that the edge can be traversed. If the graph is not directed, then it is *undirected*. A graph is said to be *connected* if for any two vertices $a$ and $b$, in the graph, there is a path $l$, consisting of a series of edges, from $a$ to $b$. A graph contains a *cycle* if it has a path that ends at the same vertex it started from. A tree is a directed graph that is connected and does not contain cycles.

For combinatorial optimization problems, decision trees can be constructed that completely enumerate all of the possible solutions to the problem. A decision tree starts with a single node, the *trunk*, and *branches* out as many times as necessary to enumerate the decisions required to solve the problem. The nodes at the extremities of the tree are the *leaves*. Except for leaf nodes, each of the nodes in the tree has one or more *child* nodes connected to it by edges directed from the *parent* node to the child node. Each set of edges and nodes connecting the leaf nodes to the trunk node represents a complete solution to the assignment problem. This means that each leaf represents a solution to the problem. The optimal solution to a combinatorial optimization problem can be obtained by completely enumerating the leaves of the tree and then finding the leaf with the optimal cost. In practice, finding the optimal solution in this manner is possible only for small-sized problems, because the associated computational complexity makes the solution time prohibitively large. It should be noted that in many practical systems obtaining a good solution in an allotted time period is more desirable than waiting until the optimal solution is reached.

For searching trees, the two basic methods are breadth-first search and depth-first search (DFS) [14]. The breadth-first search explores neighbors of a vertex, that is, outgoing edges of the vertex's predecessor in the search, before any outgoing edges of the vertex. Since the breadth-first search explores all the parent nodes before exploring the leaf nodes, it can take a long time to find a solution. Because of the desire to reach solutions quickly, the breadth-first search is not considered in this work. DFS algorithms start at the trunk of the tree and consider outgoing edges of a vertex before any neighbors of the vertex, that is, outgoing edges of the vertex's predecessor in the search. The algorithm, easily implemented with recursion, proceeds until a leaf node is reached. The selection of which node to explore at each layer is performed using heuristic arguments. Once the DFS reaches a leaf node it evaluates the next unevaluated child node of the current parent node. If there is no unevaluated node, then the algorithm checks parent nodes until there are no unevaluated nodes in the graph. In this manner the DFS systematically searches the entire tree.

One variation of the DFS is the best-first search (BFS), in which a measure is defined and used to search the tree [6]. For each parent node all the child nodes are evaluated and the best-cost node is selected. Since at each node all the children nodes need to be evaluated, the BFS is computationally intensive but tends to arrive early at good solutions. Many other search methods such as iterative broadening, limited discrepancy search, and best-leaf first search have been developed to take advantage of different types of knowledge on the system in order to quickly obtain good solutions [10]. The question of which type of search to employ depends on the requirements of the system and the ability to predict which branches will produce better solutions.

### 4.2.2 Tree Representation of WASM Scenario

Fig. 4.2 is an example of a tree in a scenario consisting of just two vehicles performing two tasks on two targets. Note that this figure shows four subtrees. The top node of each of these subtrees is connected to the root node of the entire tree. Each node represents one assignment of a vehicle to a task where the notations C$ij$ and A$ij$ denote that vehicle $i$ performs on target $j$ classification or attack, respectively.

This tree not only spans the decision space of the problem, but it also incorporates the state of the problem in its nodes. The tree is constructed by generating nodes that represent the assignment of a vehicle $i \in V$ to a task $k \in M$ on a target $j \in T$ at a specific time. The child nodes are found by enumerating all the possible assignments that can be made, based on the remaining tasks and requirements of the WASM scenario. Nodes are constructed until all combinations of vehicles, targets, and tasks have been taken into account. Note that a branch of the tree, from a root node to a leaf node, represents a feasible set of assignments for the UAV group.

In the case of the WASM problem, the tree is wide and shallow. The depth of the tree, i.e., the number of nodes from the root node to the leaf nodes, is equal to

$$N_c = N_t N_m, \tag{4.32}$$

and in the investigated problem $N_m = 3$ (the number of tasks). Since, as noted above, traversing the tree from a root node to a leaf node produces a feasible assignment, it is possible to compute such an assignment in a known time, which is the node processing rate times $N_c$.

**Figure 4.2:** WASM scenario tree, $N_v = N_t = N_m = 2$.

### 4.2.3 Search Algorithm

The proposed tree search algorithm is initiated by a BFS algorithm that provides an immediate feasible assignment. The algorithm starts at the trunk of the tree and the estimated cost of each child is calculated by using the lower bound Euclidean distance between the assigned vehicle and its designated target. The child node with the smallest estimate is selected and the cost of a flyable path to perform the assignment is evaluated using the computationally intensive, path optimization function of the MultiUAV2 simulation [9] (Appendix A) that is based on the optimization procedure discussed in Appendix B.4. The process is repeated in each layer of the tree until a leaf node is reached. The feasible assignment encoded in the branch specified by this leaf serves as a candidate optimal assignment for our tree search algorithm.

The key to an efficient tree search algorithm is its ability to quickly prune parts of the tree. The cost of a candidate optimal solution, obtained initially using the BFS algorithm discussed above, provides an upper bound on the cost of the cooperative assignment. The flowchart depicting the optimization algorithm for the WASM scenario is shown in Fig. 4.3. We begin the search at the leaf node obtained from the BFS algorithm. Using Euclidean distances we quickly obtain a lower bound on the cost of the same target being serviced by a different UAV in the group. If the cost is larger than that of the candidate optimal assignment, then this node is pruned. If not, then the computationally intensive path optimization subroutine is used to obtain the actual cost of the node with the smallest Euclidean distance, and the node is considered evaluated. We compare the cost to that of the candidate optimal assignment; if it is lower, it becomes the new candidate optimal solution. We apply these upper and lower bounding procedures iteratively traversing the tree upward and downward. This enables efficient pruning of the solution tree and obtaining a monotonically improving assignment solution. The search is terminated when all nodes have been either evaluated or pruned.

## 4.3 Genetic Algorithm

Stochastic and nongradient search methods [15] might be considered to avoid the computational complexity of the combinatorial optimization methods described in the previous two subsections and thus speed up the convergence to a good feasible solution. The GA is such an approach that does not require explicit computation of the gradients in the problem [5]. A GA will often quickly yield good feasible solutions and will not be trapped in any local minimum; however, it may not yield the optimal solution. Another key attribute of the method is the possibility of parallel implementation.

### 4.3.1 Brief Review

GAs are described in many papers and textbooks, including [2, 5], and are only briefly reviewed in this section. They enable efficient search of a decision state space of possible solutions, as long as the search space is not extremely rugged. The method involves iteratively manipulating a population of solutions, called chromosomes, to obtain a population that includes better solutions. The encoding of the GA chromosome is a major part of the solution process. After that stage has been overcome, the algorithm consists of the following steps: (1) Initialization—generation of an initial population; (2) Fitness—evaluation of the fitness of each chromosome in the population; (3) Test—stopping if an end condition is satisfied and continuing if not; (4) Find new solutions—creating new candidate chromosomes by applying genetic operators, thus simulating the evolution process; (5) Replacement—replacement of old chromosomes by new ones; and (6) Loop—going back to step 2.

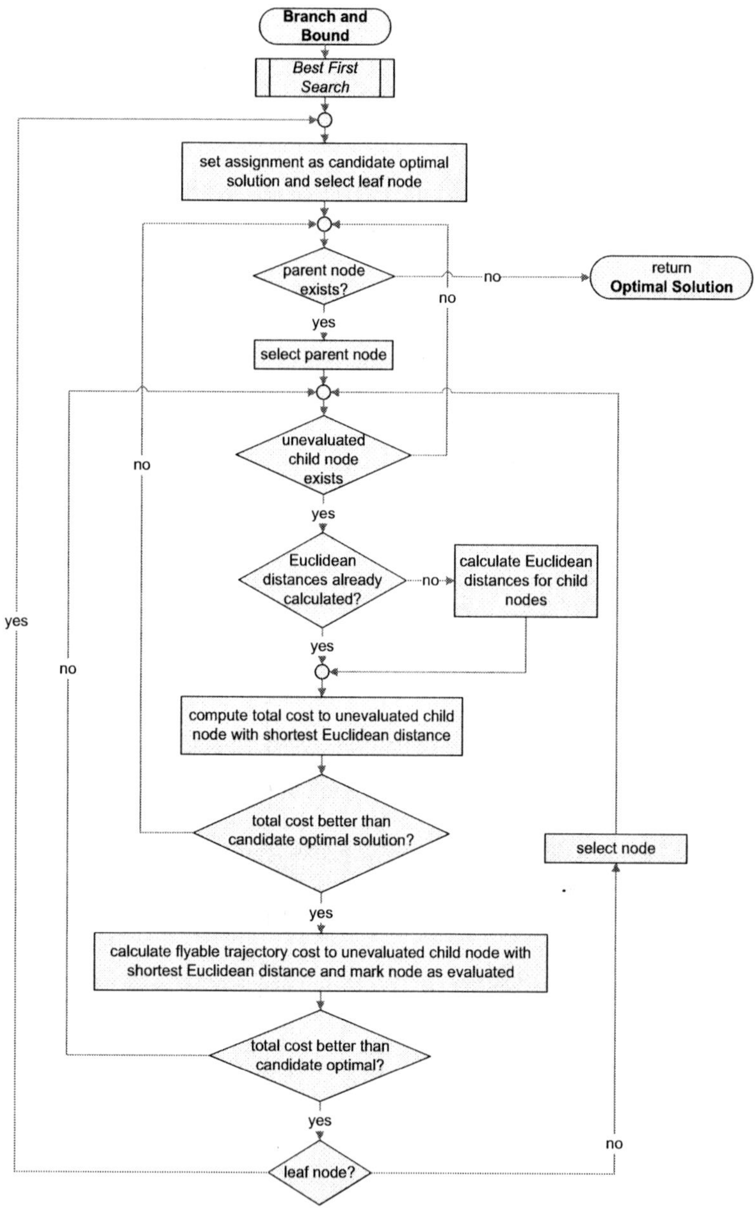


**Figure 4.3:** Tree search flowchart.

The genetic operators mentioned above are selection, crossover, mutation, and elitism. These operators are performed on the chromosome solutions consisting each of $N_c$ genes. In the selection stage two parent chromosomes are chosen from the population based on their fitness. Several selection methods are commonly used, including roulette wheel, rank, and binary tournament. In all of these methods, the better the fitness, the better the chance of

**Table 4.2:** Example of chromosome representation.

| | | | | | | | | | |
|---|---|---|---|---|---|---|---|---|---|
| Vehicle | 1 | 1 | 2 | 2 | 1 | 2 | 2 | 2 | 1 |
| Target | 2 | 3 | 1 | 3 | 1 | 1 | 2 | 2 | 3 |

being selected. Crossover is performed in single or multiple points across the chromosome, the location of which is selected randomly. For example, in the simple one-point crossover operator on chromosomes with $N_c$ genes, the child solution consists of the first $g$ genes from the first parent and $N_c - g$ genes from the second parent, and vice versa for the second child. The mutation operator involves randomly changing one of the genes in the child chromosome. This operator reduces the probability of the algorithm getting trapped in any local minimum. The elitism operator is used to save the best solutions of the generation.

### 4.3.2 GA Synthesis

The GA described here is based on the one developed in [13].

#### Encoding

The most critical part of deriving a GA is the chromosome encoding. For this problem the encoding of the GA chromosomes is based on the WASM scenario tree. It is customary to use a string for the chromosome encoding. However, for simplifying the encoding process and the application of the genetic operators, we use a matrix for our problem. Each chromosome matrix is composed of two rows and $N_c$ columns; see Table 4.2 for an example. The columns of the chromosome matrix correspond to genes. We define the set of genes as $G = \{1, 2, \ldots, N_c\}$. The first row presents the assignment of vehicle $i \in V$ to perform a task on target $j \in T$ appearing in the second row. The ordering of appearance of the target determines which task $k \in M$ is being performed; e.g., the first time a target appears in the second row (from left to right) means it has been assigned to be classified. Thus, we avoid the need to explicitly specify the task being performed and are able to simplify significantly the chromosome encoding of the assignments. We denote a feasible chromosome as defining an assignment for the cooperating vehicles performing exactly $N_m$ tasks on each of the $N_t$ targets.

An example chromosome for a problem of two vehicles ($N_v = 2$) performing the three tasks ($N_m = 3$) on three targets ($N_t = 3$) is shown in Fig. 4.2. It can be seen that the chromosome is of length $N_c = 9$ and the assignment is as follows: Vehicle 1 classifies Target 2 and then classifies Target 3. Meanwhile, Vehicle 2 classifies Target 1 and then attacks Target 3 (after it has been classified by Vehicle 1). After Target 1 has been classified (by Vehicle 2) it is attacked by Vehicle 1 and then verified by Vehicle 2. Vehicle 2 then attacks Target 2 and verifies its kill. In the meantime, Vehicle 1 verifies the kill of Target 3.

#### Generations of Solutions

The initial population is obtained by employing a random search method over the tree presented in the previous section. Producing offspring chromosomes from the parent ones, in each of the next generations, is composed of three stages: selection, crossover, and mutation. We produce an entire generation at each step and keep a constant population size. To avoid the possibility of losing the best solutions, when propagating to the new generation, we employ the *elitism* genetic operator and keep the best $N_e$ chromosomes from the previous generation. The rest of the new chromosomes ($N_s - N_e$) are obtained by repeatedly producing children until the new population is filled. At this stage the new population replaces the entire previous generation of chromosomes. Offline, the process of producing new generations can be repeated until some stopping criterion is met. For an online implementation the algorithm can be run for a specific allotted time. In this study the generations have been progressed for a given number of times, denoted $N_g$. For the selection stage, of two parents at a time, the roulette wheel method was used. On each set of two parent chromosomes we apply the crossover operator with a high probability $p_c$. Thus, there is a high probability that the parents will reproduce between themselves and a low probability that the offspring will be exact replicas of their parents.

In this study the single-point crossover method has been chosen. This point is selected randomly based on a uniform distribution. When applying this operator, the first child solution consists of the first $g \in G$ genes (columns) from the first parent and $N_c - g$ genes from the second parent, and vice versa for the second child. The application of this operator actually utilizes the implicit gradients in the problem. A crossover at one of the genes corresponds to a perturbation in the direction encoded in the other parent's gene. Note that, generally speaking, a crossover at one of the first genes corresponds to a larger perturbation than in one of the last genes since it has a higher probability of affecting the rest of the group plan.

For the children chromosome solutions to be feasible, each target $j \in T$ has to appear exactly $N_m$ times in the second row. We perform a check on the number of times each target appears in the $N_c - g$ genes switched between the parents. We start checking the second row of the switched part from its beginning (from left to right), and whenever a target appears more times than is required than that gene is changed randomly so as to reference a target that does not appear the required number of times. Thus, we enforce that each child chromosome is feasible. Note that in using this crossover mechanism we do not change the first row of the switched part, and hence the ordering of the vehicles performing tasks remains the same.

Last, the mutation operator is applied with a small probability $p_m$ to each gene (column) of the chromosome. We mutate only the identity of the vehicle $i_{old} \in V$ (first row) performing the task, so as not to affect the integrity of the assignment. The identity of the new vehicle is selected randomly such that $i_{new} \in V$ and $i_{new} \neq i_{old}$. The application of this operator prevents the algorithm from getting stuck in one branch of the tree and thus tries to avoid getting trapped in a local minimum. Generally speaking, as for the crossover operator, mutating the first genes has a larger effect on the cost than mutating the last genes, since it has a higher probability of affecting the rest of the group plan.

## 4.4 Performance Analysis

The performance of the GA, tree search, and a random search algorithm is analyzed in this section using a Monte Carlo study, consisting of 100 runs.

**Table 4.3:** Simulation parameters.

| GA | Scenario |
|---|---|
| $N_s = 200$ | $N_t \in \{1, 2, 3\}$ |
| $N_e = 6$ | $N_v \in \{1, 2, 3, 4\}$ |
| $p_m = 0.01$ | $N_m = 3$ |
| $p_c = 0.94$ | $v = 105$ m/s |
| $N_g = 100$ | $\Omega_{max} = 0.05$ rad/sec |

A comparison between the GA and MILP in the WASM was performed in [1], where it was shown that for such problems involving more than three vehicles and two targets the GA provides feasible solutions in a greatly reduced run time.

The parameters used for the simulation are summarized in Table 4.3 and the cost functions $J_1$ and $J_2$ are those given in Eqs. (4.1) and (4.3), respectively. For $J_2$ to zero weights were chosen for an early prosecution of individual targets. Note that the random search algorithm explores only feasible solutions and the best solution is kept. Two problems of different sizes are examined: 4 UAVs on 3 targets and 8 UAVs on 10 targets. The random variables are the initial location of the targets and the location and heading of the members of the UAV team.

To enable a comparison that is independent of the computer platform on which the different algorithms are run, we chose the independent variable in the following figures as the number of nodes (vehicle-target pairs) that need to be processed by the MultiUAV2 path optimization subroutine to compute the individual assignment cost. This computation, which involves optimizing the UAV trajectories for a given vehicle-target pair, is the main computational burden in running the different algorithms (between 70% and 85% in the investigated cases). For the sake of completeness the corresponding approximate run time on a personal computer (Pentium IV, 2400 MHz) is given as an additional top axis in each figure.

First, the small-size scenario of four UAVs on three targets is analyzed. In Figs. 4.4 and 4.5 the convergence of the measures of interest $J_{1min}/J_1$ and $J_{2min}/J_2$ are plotted, where $J_{1min}$ and $J_{2min}$ are the costs of the a posteriori best assignments for the cost functions of Eqs. (4.1) and (4.3), respectively (found using the tree search algorithm). For the cost function $J_1$ it is apparent that in the domain of interest of short run time (enabling onboard implementation), the GA outperforms the other methods. However, for $J_2$, while the GA still performs better than random search, the best performance is obtained using the tree algorithm. The improved performance of the tree search algorithm can be explained by noting that for the cost function $J_2$ the tree search algorithm prunes the tree considerably faster than for the cost function $J_1$, since it is dependent on the trajectory of only one vehicle. Note that on average the tree search algorithm finishes exhaustively searching the tree after $8 \times 10^5$ nodes (corresponding to a run time of approximately 114 seconds) for $J_2$ compared to $3.3 \times 10^6$ nodes (corresponding to a run time of approximately 471 seconds) for $J_1$.

Performing a complete search of the tree using the tree search algorithm in a problem of higher dimension ($N_v = 8$, $N_t = 10$) proved computationally infeasible on a personal

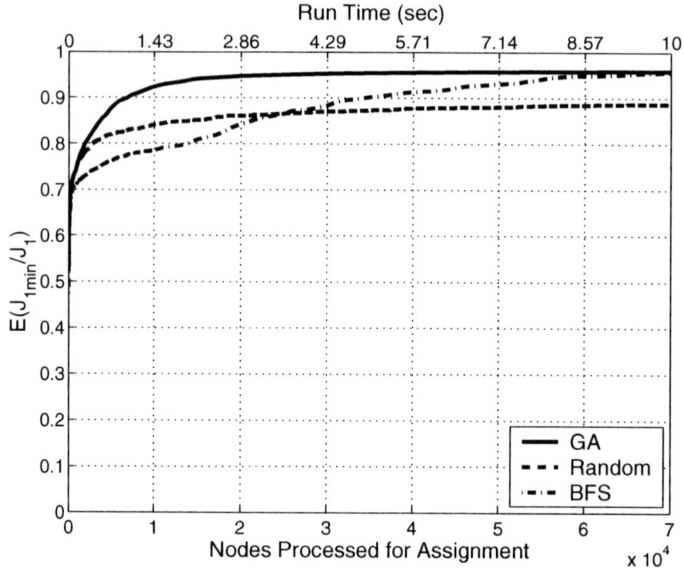


**Figure 4.4:** Mean of Monte Carlo runs: $4 \times 3 \times 3$ scenario, $J_1$ cost function.

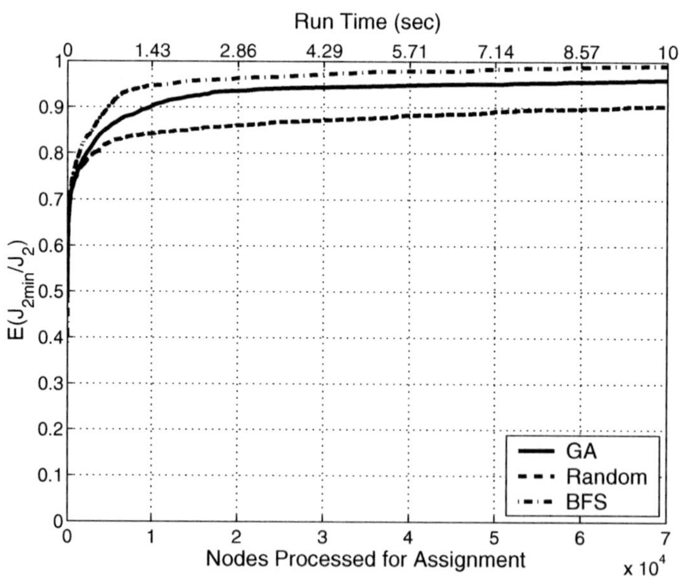


**Figure 4.5:** Mean of Monte Carlo runs: $4 \times 3 \times 3$ scenario, $J_2$ cost function.

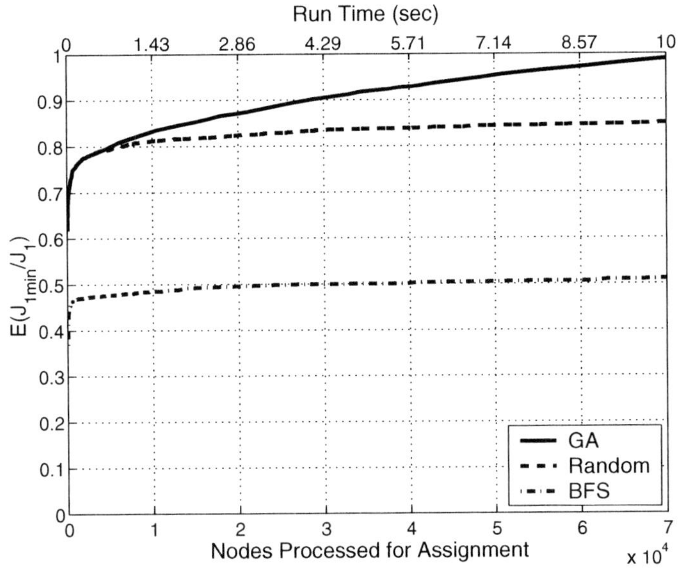


**Figure 4.6:** Mean of Monte Carlo runs: $8 \times 10 \times 3$ scenario, $J_1$ cost function.

computer (Pentium IV, 2400 MHz), since one run time of the tree search algorithm took more than three weeks. Thus, the optimal assignments were not found and the normalizing factors $J_{1min}$ and $J_{2min}$ are the costs of the best solutions found in the given time by all the methods (by GA in these cases). In Figs. 4.6 and 4.7 the average of $J_{1min}/J_1$ and $J_{2min}/J_2$ as a function of the number of nodes computed and run time is plotted for the different algorithms. The superior performance of the GA is clearly evident for both cost functions.

### 4.4.1 Algorithms Pros and Cons

The algorithms presented in this chapter have some significant advantages over the CTAP and iterative CTAP methods presented earlier. They offer the possibility of planning from the outset the entire set of task tours, instead of searching in a greedy myopic fashion. Also, timing constraints can be enforced.

In comparing the MILP, tree search, and GA methods presented in this chapter, there are two important considerations: computational complexity and path planning.

#### Computational Complexity

Incorporating all of the optimization into a single MILP results in an $NP$-hard optimization problem. Typically, nothing larger than a two-target problem can be solved in real time with three tasks per target. With only two tasks per target (for example, with classify and attack grouped into a single task, and the attack aborted if classify fails), problems involving three targets can be solved fast enough for online implementation, but larger problems cannot [11, 12]. These problem sizes are sufficient for small WASM teams of up to four vehicles;

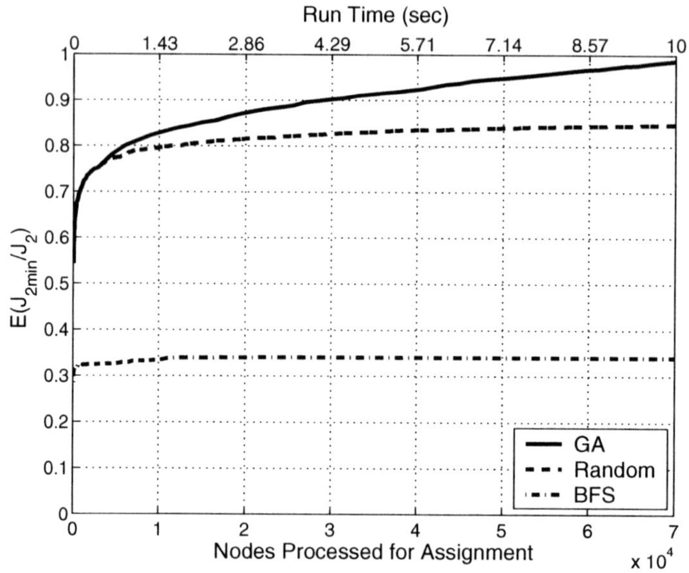


**Figure 4.7:** Mean of Monte Carlo runs: $8 \times 10 \times 3$ scenario, $J_2$ cost function.

however, scaling difficulties limit their utility for larger problem sizes. The MILP can still be used to develop optimal solutions offline for medium-sized problems.

Similarly to MILP, tree search can be used to solve online small-sized problems. For medium to large-sized problems, involving more than a handful of vehicles and targets, it may only be used offline.

In contrast, GA can be used to provide an online solution to problems of different sizes. For small-sized problems it offers no advantages over MILP and tree search, while for larger-sized problems it can provide good feasible solutions that improve with runtime.

### Path Planning

The MILP formulation presented here requires knowledge of the flight time $t_{ij}^{vk}$ required for each task. This can be calculated directly for the first task of each vehicle. However, the flight times for later tasks will depend on the angle at which the vehicle approached the previous target. The resulting flight times cannot, therefore, always be known accurately a priori. Euclidean distances between targets can be used to approximate the flight path length, and if the turn radius of the vehicles is small compared to the target spacing, this is a practical assumption. However, if the target spacing is on the order of the turn radius of the vehicles, Euclidean distances can result in very misleading results. To compensate for the errors associated with the Euclidean distance assumption, slack time can be introduced into the flight time variables $t_{in}^{vk}$. Alternately, Euclidean distance can be used in the MILP calculation, and, after an assignment has been calculated, flyable paths can be calculated, with loiter time added when necessary to ensure timing constraints are satisfied. Such heuristic fixes, unfortunately, sacrifice the optimality which is the primary advantage of using MILP.

The tree search and GA use piecewise optimal Dubins' trajectories for the path planning. Thus, the obtained trajectory, for a vehicle that has to perform more than one task, is in general not optimal. However, it should be noted that as in the GA an entire assignment is evaluated, then, theoretically, the optimal path for a vehicle performing multiple tasks may be found. We use the term "theoretically" as this process may be intractable if more than two tasks are involved.

## 4.5 Summary

In this section the multiple-tasks assignment problem associated with the WASM scenario has been solved using MILP, tree search, and GA. The problem has prohibitive computational complexity requiring making simplifying assumptions so as to be solvable using MILP. The proposed tree search algorithm has desirable qualities, such as providing a fast initial feasible solution that monotonically improves and, eventually, converges to the optimal solution. The algorithm can be applied online to small-sized problems, providing the best piecewise optimal solution, whereas for large-sized problems an immediate feasible solution that improves over run time is obtained. A GA was proposed for obtaining good feasible solutions for computationally intensive large-sized problems. It was shown that the GA can provide good solutions considerably faster than other search methods. This enables real-time implementation for high-dimensional problems.

## Bibliography

[1] M. Darrah, W. Niland, B. Stolarik, and L. E. Walp. UAV cooperative task assignments for a SEAD mission using genetic algorithms. In *Proceedings of the AIAA Guidance, Navigation, and Control Conference*, Keystone, CO, 2006.

[2] D. E. Goldberg. *Genetic Algorithms in Search, Optimization, and Machine Learning*. Addison-Wesley, Reading, MA, 1989.

[3] N. Hartsfield and G. Ringel. *Pearls in Graph Theory: A Comprehensive Introduction*. Academic Press, Boston, 1994.

[4] A. Kaufmann. *Graphs, Dynamic Programming, and Finite Games*. Mathematics in Science and Engineering 36. Academic Press, New York, 1967.

[5] M. Mitchell. *An Introduction to Genetic Algorithms*. MIT Press, Boston, MA, 1996.

[6] J. Pearl. *Heuristics: Intelligent Search Strategies for Computer Problem Solving*. Addison-Wesley, Reading, MA, 1984.

[7] M. Pinedo. *Scheduling*. Prentice–Hall, Englewood Cliffs, NJ, 2002.

[8] S. J. Rasmussen and T. Shima. Tree search for assigning cooperating UAVs to multiple tasks. *International Journal of Robust and Nonlinear Control*, 18(2):135–153, 2008.

[9] S. J. Rasmussen, J. W. Mitchell, C. Schulz, C. J. Schumacher, and P. R. Chandler. A multiple UAV simulation for researchers. In *Proceedings of the AIAA Modeling and Simulation Technologies Conference*, Austin, TX, 2003.

[10] W. Ruml. *Adaptive Tree Search*. Ph.D. thesis, Harvard University, Cambridge, MA, May 2002.

[11] C. Schumacher, P. R. Chandler, M. Pachter, and L. Pachter. Constrained optimization for UAV task assignment. In *Proceedings of the AIAA Guidance, Navigation, and Control Conference*, Providence, RI, 2004.

[12] C. Schumacher, P. R. Chandler, M. Pachter, and L. Pachter. UAV task assignment with timing constraints via mixed-integer linear programming. In *Proceedings of the AIAA Unmanned Unlimited Conference*, Chicago, IL, 2004.

[13] T. Shima, S. J. Rasmussen, A. Sparks, and K. Passino. Multiple task assignments for cooperating uninhabited aerial vehicles using genetic algorithms. *Computers and Operations Research*, 33(11):3252–3269, 2005.

[14] J. Siek, L.-Q. Lee, and A. Lumsdaine. *The Boost Graph Library: User Guide and Reference Manual*. Addison-Wesley, Boston, 2002.

[15] J. C. Spall. *Introduction to Stochastic Search and Optimization*. John Wiley & Sons, New York, 2003.

[16] P. Toth and D. Vigo. *The Vehicle Routing Problem*. SIAM, Philadelphia, 2002.

Chapter 5

# Simultaneous Multiple Assignments

*Corey Schumacher and Tal Shima*

*The whole is more than the sum of its parts.*
—Aristotle

In this chapter the problem of assigning multiple UAVs to simultaneously perform cooperative tasks on consecutive targets is treated. As in the WASM scenario, multiple tasks need to be performed on each target, but the difference is that these multiple tasks need to be performed simultaneously. This complicates the coupling between task assignment and path planning for each UAV.

## 5.1 Cooperative Moving Target Engagement

The scenario under consideration is one of cooperative moving target engagement (CMTE). In this scenario, a stand-off UAV with a wide-area ground moving target indication (GMTI) Doppler radar and several smaller, stand-in UAVs, with small-area GMTI Doppler radars and GPS-guided ground-attack weapons, must cooperatively track and prosecute moving ground targets. A very large number of potential targets could simultaneously be tracked by the stand-off vehicle, but we will assume that targets to be prosecuted are nominated to the UAV team by a human supervisor or outside agent. Thus, we focus on the cooperative task planning required to track and attack the assigned targets. To meet the potentially severe timing constraints of such a scenario, the computation of an efficient set of task assignments and corresponding vehicle paths must be automated for practical implementation.

### 5.1.1 CMTE Scenario

The CMTE requires that two or more UAVs with Doppler radars track a moving ground target while an additional UAV launches a GPS-guided munition. The sensed target position

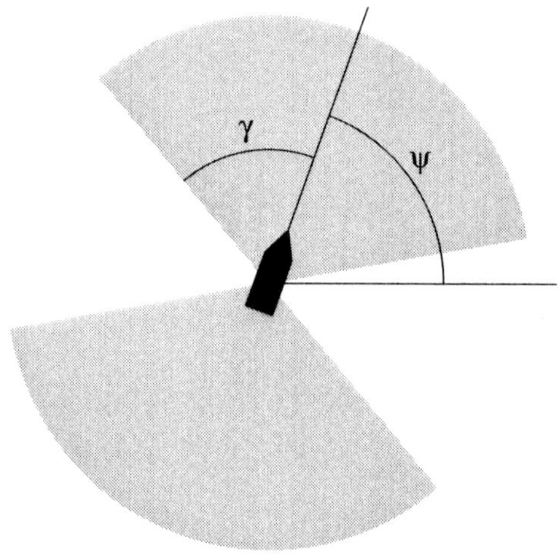


**Figure 5.1:** Region of detectability based on target heading.

and associated error ellipses from each tracking UAV are fused to form an accurate GPS location of the target, to which the munition is guided. To reduce the covariance of the error in the position estimate of the moving target, the UAVs tasked to perform the tracking must have sufficient line-of-sight (LOS) separation to the target, preferably near orthogonal views. Therefore, we require at least a 45° separation between the look angles of the two vehicles tracking a target, restricting the tracking regions and time windows. In addition, a target moving at a given speed can be tracked by the Doppler radars only if the LOS view to the target is within some offset angle, $\gamma$, from the heading of the moving target, $\psi$. Thus, the region of detectability is

$$\Omega = \{\psi - \gamma, \psi + \gamma\} \cup \{\pi + \psi - \gamma, \pi + \psi + \gamma\}. \tag{5.1}$$

Fig. 5.1 shows the heading of the target and the associated angular regions in which UAVs carrying GMTI radars can be located to detect the target's motion.

Complicating matters further, each stand-in UAV has a limited sensor footprint with minimum and maximum ranges $(r_{min}, r_{max})$ and a maximum angle away from the body y-axis $(\theta_r)$ that the radar can view. Due to the configuration of the radar antenna array, the footprint is pointed out the wing of the UAV. Fig. 5.2 shows a stand-in UAV tracking a target and the associated sensor footprint relative to the orientation of the UAV. The sensor can scan on either side of the UAV, but not both at the same time. The dark area in this figure corresponds to the sensor footprint and the light area corresponds to the target's region of detectability (plotted in Fig. 5.1).

For simplicity, we will assume that the stand-off vehicle has a wide-area coverage GMTI sensor with a 360° sensor footprint able to view the entire field. This assumption can be relaxed but requires additional timing constraints to be included in the assignment formulation to account for the time windows when the stand-off vehicle is, and is not, available to perform tracking tasks. We will also assume that the stand-off vehicle travels in a tight circular orbit, so that its LOS to targets is approximately fixed. This assumption

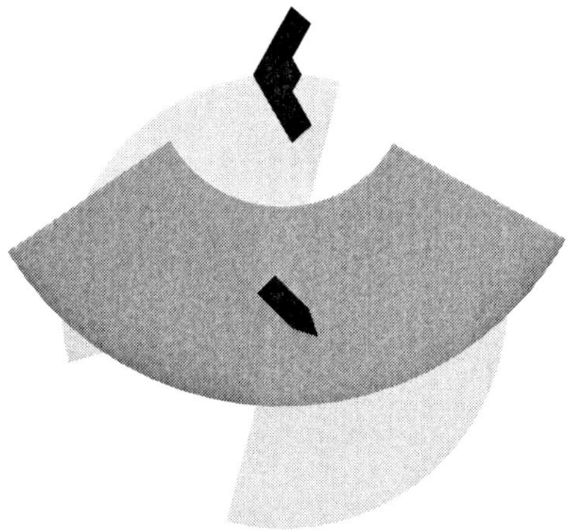

**Figure 5.2:** UAV sensor footprint (dark gray region).

can also be relaxed, with additional complications in the calculation of cooperative tracking trajectories. This vehicle can track any ground moving target within the region of interest, as long as the component of the target's velocity, relative to the terrain in the direction of the stand-off UAV, is above the required minimum detection speed $v_d$. UAVs inside the area of interest can cooperatively track a target with the off-board vehicle, or with another stand-in team member. To reduce the target location estimation error, the difference in bearing angles of the tracking UAVs to the target is required to be greater than 45°. This restriction partitions the detectability region of the target further into regions that satisfy both the target detectability requirement and the angle offset requirement. For fixed target heading and position and fixed off-board vehicle position, regions where a target can be cooperatively tracked can be identified and used to develop path-planning routines of the UAVs to complete the mission.

While the bulk of the complexity in the CMTE scenario comes from the cooperative tracking of targets, the attack on the target must also be considered. All UAVs inside the area of interest can track targets and drop weapons. To be in position to attack, a UAV must be pointed at the target and be between a minimum and maximum range. Once the weapon is launched, the attacking UAV is free to be reassigned to other tasks, but the UAVs tracking the target must track the target for the duration of the weapon flight.

To significantly reduce the position uncertainty each target must be simultaneously tracked by two vehicles with sufficient angular separation. Thus, during an attack, a target inside the region of interest is tracked by a stand-in vehicle cooperating with either a stand-off vehicle or another stand-in UAV. Let

$$\mathcal{V} = \{0, 1, \ldots, V\} \tag{5.2}$$

be the set of cooperating UAVs where 0 denotes the unarmed stand-off UAV and there are $V$ stand-in UAVs, indexed by $v = 1, \ldots, V$, carrying GPS-guided munitions. We assume that the vehicles fly at a constant speed and that fuel constraints are not a factor during a task assignment. Let

$$\mathcal{G} = \{1, \dots, N\} \tag{5.3}$$

be the set of ground targets nominated to the UAV team for prosecution. This set of targets must be serviced by the $V + 1$ cooperating UAVs. For simplicity we assume that throughout the engagement the targets have a constant velocity, i.e., constant speed and heading. Let

$$\mathcal{K} = \{1, \dots, K\} \tag{5.4}$$

be the set of tasks indexed $k = 1, \dots, K$ to be performed on each target. Let

$$\mathcal{S} = \{1, 2, \dots, S\} \tag{5.5}$$

be the set of stages in which the assignment of three vehicles to each target is made. Since each target need be serviced by a UAV trio only once, then $S = N$. Note that it is possible for a task at a late stage to be completed prior to that of a preceding one. This may happen if different stand-in vehicles participate in these phases of the assignment.

### 5.1.2 Simultaneous Tasks

This CMTE scenario features a highly constrained task assignment and scheduling problem, requiring close coordination of task execution times and durations. Substantial constraints are imposed by vehicle dynamics, target motion, sensor capabilities, and weapon characteristics. CMTE also features complications that did not exist in the scenarios presented in Chapters 3 and 4. Specifically, CMTE includes simultaneous or "joint" tasks, overlapping tasks, and task durations. Two or more vehicles must track simultaneously, or no benefit is gained. The tracking tasks must continue for the entire duration of the weapon flight, or the mission fails. Due to the complexity of this scenario, methods such as the capacitated transshipment problem formulation used in Chapter 3 can be applied only with extensive simplifying assumptions that greatly restrict the potential solution space. This complexity makes the CMTE scenario an excellent problem for studying time-dependent cooperative assignment and scheduling methods. This chapter presents methods applicable to this more complex class of problems, involving time availability windows of resources, joint and overlapping tasks, and task durations. Other UAV cooperative control scenarios featuring joint and overlapping tasks, such as Suppression of Enemy Air Defenses (SEAD) missions involving jamming, can be approached using similar methods [1].

In total, each target $n \in \mathcal{G}$ requires three overlapping tasks: two simultaneous tracking tasks, from two separate UAVs, and an attack task, performed by a third stand-in UAV. The simultaneous cooperative tracking tasks must begin before the weapon is launched and continue until the weapon strikes the target. Thus, the tracking duration $\Delta T^n$ must satisfy

$$\Delta T^n \geq t_{flight}, \tag{5.6}$$

where $t_{flight}$ is the flight time of the weapon. For simplicity we assume that weapons are launched at a fixed distance from the target, resulting in a weapon flight time, $t_{flight}$, that is known and fixed.

## 5.2 Combinatorial Optimization Problem

### 5.2.1 Time Availability Windows

These numerous dynamic and sensor constraints, as well as the complexities of integrating complex task assignment and path planning components, are abstracted out of the task assignment and scheduling problem using a "time availability window" approach. This approach results in a three-stage process for the full planning problem. First, the path planning is abstracted into time availability windows in which a task can be performed. All potential task execution times are reduced to sets of time windows. Second, the optimization problem is then formulated and solved using these time availability windows as a mathematical abstraction of the path planning, resulting in specific task assignments and task schedules. Finally, specific flight paths are calculated to meet the assigned task schedule.

Path optimization for the stand-in UAVs in the CMTE is subject to numerous constraints. For this work, we simplify the constraints by assuming that relative to the UAVs the targets are approximately fixed in position. We also restrict the cooperative tracking tasks to using a circular orbit with a prespecified radius $r_t$, where $r_{min} \leq r_t \leq r_{max}$. This path-planning strategy is suboptimal but allows sufficient freedom to study the task assignment and scheduling problem.

### 5.2.2 Path Planning

Dubins' car [2], discussed in Appendix B, is a classic model for representing the dynamics of carlike robots and unmanned aerial vehicles traversing planar paths with bounded curvature without direction reversal. Thus, although six degree-of-freedom dynamics are used in the vehicle simulation, for the purposes of path planning the UAV dynamics are modeled as a Dubins' car.

Path planning is performed in several steps:

1. Calculate the set of all feasible times $T^n_{v1,v2,v3}$ that vehicles $v1, v2, v3 \in V$ ($v1 \neq v2 \neq v3$) can begin performing the specified tasks on target $n \in \mathcal{G}$ for a duration of $\Delta T^n$. This involves calculating the minimum time at which a vehicle can begin tracking the target and adding loiter to the path before tracking begins if a delay is needed to satisfy timing constraints. For most vehicle and target initial conditions, any task start time that is later than the minimum one is feasible. However, due to the dynamic constraints of the vehicle in some special cases continuous elongation of the vehicle path is not feasible [5]. Determining the feasible time windows does not require calculation of a path for every possible start time.

2. Select the combination of time windows that allows minimum task completion time. Once the set of all possible times $T^n_{v1,v2,v3}$ that vehicles $v1, v2, v3 \in \mathcal{V}$ can perform the required tasks on target $n \in \mathcal{G}$ have been calculated, we select the combination resulting in the earliest possible completion of target prosecution, while satisfying all constraints. Let $t^{l,n}_{v1,v2,v3}$ denote the time the task, performed by a UAV trio, on target $n$ at stage $l \in \mathcal{S}$ is completed. For the fused estimate of the target position to have sufficient accuracy, the look angles of the tracking vehicles must be substantially separated.

3. Calculate paths that meet the time windows specified in the previous step. Once specific start and stop times for each task are known, paths are calculated to meet those exact timing constraints. The vehicles' positions at the end of these tasks, and the completion times, are saved and used as the new initial conditions for calculating vehicle paths for succeeding tasks.

Further details of the Dubins' car dynamic model and the methods used for path planning calculation can be found in Appendix B.

## 5.3 Optimization via MILP

In many cooperative scenarios, the ability to assign cooperative tasks (i.e., two or more agents are assigned subtasks at the same target) is critical to mission performance. This assignment is often complicated by the fact that agents may have windows in time when they can complete specific tasks. In a UAV scenario, these windows are typically a product of the underlying dynamic constraints of the vehicle. For example, if a coordinated task requires one agent to track a target while the other attacks it, then one agent may need to adjust its path to wait for its partner. If the path cannot be extended smoothly (e.g., UAV attack paths cannot be extended without discontinuity in some cases [4]), then there are separate intervals in which the task can be done with a period of infeasibility in between. A MILP formulation of the problem allows these constraints to be addressed. Once the problem nomenclature has been posed, linear constraints are given that satisfy the requirements of time-dependent task assignment.

### 5.3.1 Nomenclature

Let $k$ be the number of tasks to be performed on each target, $\mathcal{V}$ be the number of agents, and $V$ be the number of targets. Also, let $x^{n\,(w)}_{v,k}$ be a binary decision variable indicating that agent $v$ is to perform task $k$ at target $n$ starting in time window $(w)$. Note that the number of decision variables grows as the product of the numbers of agents, targets, tasks, and time windows. To allow for the case in which there are more agents than tasks to be done, let $z_v$ be a binary decision variable indicating that agent $v$ is to be assigned the null task (assumed to have zero cost). The case in which there are more tasks than agents must be addressed in an iterative manner and will be discussed in section 5.3.3. To enforce the timing between tasks and to optimize the total time of the assignment, let $t^k_n$ be the time that task $k$ is started on target $n$. For the WASM scenario discussed earlier, each task is assumed to occupy only an instant in time, so that the time a vehicle begins the task is also the time the task is completed. In the CMTE scenario, tracking tasks have a duration, and $t^k_n$, the time a task is started, can be different from the time a task is completed.

To represent the cost associated with each of the decision variables, the windows of time when agent $v$ can accomplish task $k$ at target $n$ must be calculated. For each $(v, k, n)$, a set $W^n_{v,k}$ is formed where each element of $W^n_{v,k}$ contains a start time $(T^{n\,\lfloor w}_{v,k})$ of window $w$ and a stop time $(T^{n\,w\rfloor}_{v,k})$. This can be done in *any* manner suitable to the problem definition. It is this flexibility that gives this assignment method its appeal—as long as the underlying path planning can be represented by windows of task availability, this assignment algorithm can be used. Note that a worst-case target prosecution time exists for each target and the maximum of those times is the upper bound on the final task completion time. Practically, an upper bound on target completion time will always exist due to fuel constraints. This maximum target completion time, $T_0$, allows the formation of timing inequalities that support task precedence.

### 5.3.2 Constraints

A MILP is defined by linear constraints and the cost function to be optimized. In a manner similar to the development in section 4.1, a set of constraints and an associated cost function to accomplish the assignment can be developed.

#### Nontiming Constraints

1. Each agent is assigned exactly one task or goes to the sink.

$$\sum_{n=1}^{N}\sum_{k=1}^{K}\sum_{w} x_{v,k}^{n\,(w)} + z_v = 1 \tag{5.7}$$
$$\text{for } v = 1, \ldots, V.$$

2. Any target that receives one task receives all tasks.

$$\sum_{v=1}^{V}\sum_{w} x_{v,1}^{n\,(w)} = \sum_{v=1}^{V}\sum_{w} x_{v,k}^{n\,(w)} \tag{5.8}$$
$$\text{for } n = 1, \ldots, N, k = 2, \ldots, K.$$

3. Each target is serviced at most once. In combination with the above constraint, this also ensures that each target receives each task at most once.

$$\sum_{v=1}^{V}\sum_{w} x_{v,1}^{n\,(w)} \leq 1 \tag{5.9}$$
$$\text{for } n = 1, \ldots, N.$$

#### Timing Constraints

1. The time to start task $k$ at target $n$ must be in window $w$ of agent $v$ if the corresponding decision has been made. Note that these inequalities are trivially satisfied when $x_{v,k}^{n\,(w)}$ is zero but become tight restrictions on $t_n^k$ when $x_{v,k}^{n\,(w)}$ is one.

$$t_n^k \geq T_{v,k}^{n\,\lfloor w} - (1 - x_{v,k}^{n\,(w)})2NT, \tag{5.10}$$
$$t_n^k \leq T_{v,k}^{n\,w\rfloor} + (1 - x_{v,k}^{n\,(w)})2NT \tag{5.11}$$
$$\text{for } k = 1, \ldots, K, v = 1, \ldots, V, n = 1, \ldots, N, w \subset W_{v,k}^{n}.$$

2. If precedence of tasks is required, then constraints similar to the following will be needed. Here we constrain the tasks to occur in order $1, \ldots, K$.

$$t_n^k \le t_n^{k+1} \quad \text{for } n = 1, \ldots, N, k = 1, \ldots, K-1. \tag{5.12}$$

To have agents cooperate in servicing a target simply requires defining the relative timing of tasks. If, for example, two UAVs were to start two cooperative tasks (say, task 1 and 2) simultaneously, then the constraint $t_n^1 = t_n^2$ could be added. Similarly, if a task must occur within some interval after a previous task is performed, a constraint pair like $(t_n^2 \le t_n^1 + \delta t_n^1, t_n^2 \ge t_n^1)$, where $\delta t_n^1$ is the desired maximum task interval, is applied.

3. To ensure that as many targets as possible are serviced, we impose a missed target penalty. Note that all $t_n^k$ that are associated with targets that are missed are equal to $MT_0$, where $M$ is the number of missed targets. This constraint also eliminates the degenerate solution of assigning all agents to the null task.

$$t_n^k \ge \left(\sum_{m=1}^{N} \left\{1 - \sum_{v=1}^{V} \sum_{w} x_{v,k}^{m\,(w)}\right\}\right) T - \left(\sum_{v=1}^{V} \sum_{w} x_{v,k}^{n\,(w)}\right) 2NT, \tag{5.13}$$

$$t_n^k \le \left(\sum_{m=1}^{N} \left\{1 - \sum_{v=1}^{V} \sum_{W} x_{v,k}^{n\,(w)}\right\}\right) T + \left(\sum_{v=1}^{V} \sum_{w} x_{v,k}^{n\,(w)}\right) 2NT \tag{5.14}$$

for $k = 1, \ldots, K, j = 1, \ldots, n$.

### 5.3.3 Applying MILP

To reduce the CMTE scenario to a more reasonable level, the following assumptions and restrictions will be added:

1. Targets have constant heading.
2. Targets have fixed position. (Since the UAVs have large sensor footprints relative to the distance a target could travel in typical scenarios, this is a reasonable assumption.)
3. Tracking of targets occurs along arcs of a circle centered at the target with radius so as to place the target in the center of the sensor footprint (see Fig. 5.2).
4. Weapons are launched at a fixed distance from the target and flight time of the weapon is known so as to fix the amount of time after an attack has occurred that the target must be tracked.

These restrictions and assumptions simplify the level of path planning needed to accomplish a CMTE mission. Additional complexity could be added without changing the method of assignment, as long as the interface between the nonlinear path planning and the assignment algorithm remains abstracted as time availability windows for the team agents.

Because the CMTE scenario hinges on the ability of UAVs to cooperatively track a moving target, much of the assignment complexity is involved with determining which team members are assigned to track which target and with whom. To this end, the basic time-dependent assignment algorithm developed in section 5.3.2 is augmented with additional decision variables to allow pairwise decisions. Let $y_{u,v}^{n\,(w)}$ be a binary decision variable indicating that UAVs $u$ and $v$ are assigned to cooperatively track target $n$ in time window $w$. This allows the path planning routines to calculate windows of time when the pair of vehicles $(u, v)$ can cooperatively track a target.

Following the nomenclature established above and in section 5.3, let $k = 1$ be designated the *attack* task and let $k = 2$ be the *track* task; then the following constraints are used to assign a team of UAVs in the CMTE scenario.

### Nontiming Constraints

1. Each agent is assigned exactly one task or goes to the sink. An agent can be assigned to cooperatively track a target with the off-board vehicle ($x_{v,2}^{n\,(w)}$) or with another inside team member ($y_{u,v}^{n\,(w)}$), but not both. At a higher level, an agent could be assigned to attack ($x_{v,1}^{n\,(w)}$) or track, but not both.

$$\sum_{n=1}^{N}\sum_{k=1}^{K}\sum_{w} x_{v,k}^{n\,(w)} + \sum_{n=1}^{N}\sum_{u=1,u\neq v}^{V}\sum_{w} y_{u,v}^{n\,(w)} + z_v = 1$$
$$\text{for } v = 1, \ldots, V. \tag{5.15}$$

2. Any target that receives one task receives all tasks. Since the *track* task occurs in two different decision variables, the sum of both must equal the decision to complete the *attack* task by another vehicle.

$$\sum_{v=1}^{V}\sum_{w} x_{v,1}^{n\,(w)} = \sum_{v=1}^{V}\sum_{w} x_{v,2}^{n\,(w)} + \sum_{u=1}^{v-1}\sum_{v=u+1}^{V}\sum_{w} y_{u,v}^{n\,(w)}$$
$$\text{for } n = 1, \ldots, N. \tag{5.16}$$

3. Each target is serviced at most once. In combination with the above constraint, this also ensures that each target receives each task at most once.

$$\sum_{v=1}^{V}\sum_{w} x_{v,1}^{n\,(w)} \leq 1 \tag{5.17}$$
$$\text{for } n = 1, \ldots, N.$$

### Timing Constraints

The CMTE scenario requires that a target be continuously tracked for the duration of the weapon flight after the weapon has been launched. Since it is assumed that the weapon is launched from a specific distance from the target and the speed of the munition is known, the path planning routines can return windows of time when an agent (or pair of agents) can start tracking a target and continue for at least $t_\alpha$, the time from munition launch to impact. This also allows the compaction of $t_n^1$ and $t_n^2$ to $t_n$ since we can constrain the tracking to begin when the weapon is launched without changing the requirements of the scenario.

1. Time to prosecute target $n$ must be in window $w$ of agent $v$ if the corresponding decision has been made.

$$t_n \geq T_{v,k}^{n\,\lfloor w} - (1 - x_{v,k}^{n\,(w)})2NT, \tag{5.18}$$

$$t_n \leq T_{v,k}^{n\,w\rfloor} + (1 - x_{v,k}^{n\,(w)})2NT, \tag{5.19}$$

$$\text{for } k = 1, \ldots, K,\ v = 1, \ldots, V,\ n = 1, \ldots, N,\ w \subset W_{v,k}^n,$$

$$t_n \geq T_{q,v}^{n\,\lfloor w} - (1 - y_{u,v}^{n\,(w)})2NT, \tag{5.20}$$

$$t_n \leq T_{q,v}^{n\,w\rfloor} + (1 - y_{u,v}^{n\,(w)})2NT, \tag{5.21}$$

$$\text{for } u = 1, \ldots, V{-}1,\ v = u{+}1, \ldots, V,$$
$$n = 1, \ldots, N,\ w \subset W_{u,v}^n.$$

2. Due to the reduction of timing variables to the representative target prosecution time, no additional task timing constraints are needed.

3. A missed target penalty carries over from Eqs. (5.13) and (5.14) to ensure that all UAVs are not assigned to the null task. Note that only $x_{v,1}^{n\,(w)}$ needs to be examined due to the coupling through constraint (5.16), which ensures that targets that are not attacked also will not be tracked.

$$t_n \geq \left(\sum_{m=1}^{N}\left\{1 - \sum_{v=1}^{V}\sum_{w} x_{v,1}^{m\,(w)}\right\}\right)T - \left(\sum_{v=1}^{V}\sum_{w} x_{v,1}^{n\,(w)}\right)2NT, \tag{5.22}$$

$$t_n \leq \left(\sum_{m=1}^{N}\left\{1 - \sum_{v=1}^{V}\sum_{w} x_{v,1}^{m\,(w)}\right\}\right)T + \left(\sum_{v=1}^{V}\sum_{w} x_{v,1}^{n\,(w)}\right)2NT \tag{5.23}$$

$$\text{for } n = 1, \ldots, N.$$

### Cost Function

The CMTE scenario requires that all targets are prosecuted as quickly as possible. As with the WASM scenario discussed in section 4.1, a variety of cost functions are possible. A good choice is to minimize the sum of the total target prosecution times:

$$J = \sum_{n=1}^{N} t_n. \tag{5.24}$$

Eqs. (5.15)–(5.24) define a MILP suitable for assigning each UAV in a team to one task in a CMTE scenario. Since there may not be enough UAVs to fully prosecute all available targets without multiple tasks per UAV, the assignment algorithm must be iterated to produce a tour of assignments. A truly optimal solution would require the optimization of path planning and assignment to be coupled. By applying this assignment scheme with the path planning constraints specified earlier, a tour of assignments will be suboptimal, but the reduced computational complexity allowed by separating the path planning and the assignment is necessary to achieve reasonable computation times.

The time windows calculated in the underlying path planning routines can easily be shifted by the amount of time a vehicle has already committed to, so an iteration can be used where the state (position and time committed) of the vehicles is updated after each assignment stage is run. Once the assignment algorithm has allocated a task to each vehicle in the team, the earliest target prosecution time is selected. The target corresponding to that time is removed from the target list and the vehicles assigned to the tasks pertaining to the prosecution of that target have their states updated to reflect the time it will take to complete their respective tasks. Vehicles associated with tasks related to the prosecution of other targets are not updated. New time windows are computed with the updated positions and shifted by the amount of time committed to earlier stages. The assignment algorithm is iterated until all targets have been prosecuted. The MILP procedure presented here is thus, in some ways, a synthesis of the iterative CTAP and MILP strategies presented earlier for the WASM problem. Solving the global optimal MILP for all targets simultaneously, without the iterative simplification, cannot be done quickly enough for real-time implementation. However, an alternative to an iterative solution is to use suboptimal solutions of the global MILP, which can be generated more rapidly by MILP solution software such as CPLEX [1]. Monte Carlo analysis would be needed to determine what was the best option, on average, for a specific scenario.

Obtaining a solution to any MILP formulation is *NP*-hard. This assignment algorithm is based on MILP and thus is also *NP*-hard, resulting in extreme amounts of computation needed for large problems. A number of different random CMTE situations were simulated to estimate the computation required for various problem sizes. It was found that problems in which the sum of the number of vehicles and targets is less than or equal to 12 are solvable in less than a minute on a modern desktop computer [3]. For many CMTE missions, small problem sizes are typical, involving five UAVs and three to four targets. Larger problems will require a reformulation of the assignment algorithm or a partitioning of the team and target space into problems small enough to be solved in a timely manner. Another option would be to run the MILP solver for a limited period of time and then use the best available solution.

## 5.4 Genetic Algorithm

In this section a GA is presented for the CMTE scenario. The algorithm, derived in [6], is somewhat similar to the one discussed in the previous chapter.

### Encoding

Let

$$C = \{1, 2, \dots, N_c\} \tag{5.25}$$

be the set of each chromosome's genes, where $N_c = 3N_t$. Each chromosome matrix is composed of two rows and $N_c$ columns corresponding to genes. In the first row the vehicles performing the tasks are defined, while in the second row the serviced target is denoted.

**Table 5.1:** Example of chromosome representation.

| Vehicle | 5 | 4 | 2 | 0 | 2 | 4 | 3 | 0 | 1 | 0 | 2 | 1 |
|---|---|---|---|---|---|---|---|---|---|---|---|---|
| Target | 2 | 2 | 2 | 3 | 3 | 3 | 1 | 1 | 1 | 4 | 4 | 4 |

Every set of three genes defines a simultaneous task on one target. The first two of three genes in each set define the vehicles performing the cooperative tracking, while the third gene defines the vehicle performing the attack by releasing a weapon.

An example chromosome is given in Table 5.1 where six vehicles denoted 0 to 5 (appearing in the first row of the chromosome) perform tasks on four targets denoted 1 to 4 (appearing in the second row of the chromosome). The encoded assignment is as follows:

- Vehicles 4 and 5 are assigned to track Target 2 with Vehicle 2 assigned to attack it.
- Vehicle 2 together with the stand-off UAV (denoted as vehicle 0) are assigned to track Target 3 with Vehicle 4 assigned to attack it. Since Vehicle 2 services also Target 2, then the task on Target 3 can be performed only after that on Target 2 has been accomplished.
- The stand-off UAV and Vehicle 3 are assigned to track Target 1 with Vehicle 1 assigned to attack it. Although Vehicle 0 is assigned to both Targets 1 and 3, these tasks can be performed independently as this stand-off UAV can track multiple targets simultaneously
- Vehicle 2 and the stand-off UAV are assigned to track Target 4 with Vehicle 1 assigned to attack it. Since Vehicle 1 services also Target 1, then the task on Target 4 can be performed only after that on Target 1 has been accomplished.

## Genetic Operators

We initialize the algorithm with $N_f$ chromosomes and maintain a constant size solution population throughout the different stages. Offline, the process of producing new generations can be repeated until some stopping criterion is met; e.g., the fitness of the best chromosome hasn't improved for a given number of iterations. For an online implementation the algorithm can be run for a specific allotted time. In this study the population has been processed for a given number of generations, denoted $N_g$. For creating a new population of solutions we employ the genetic operators: selection, crossover, mutation, and elitism.

### Selection

In the selection stage two parent chromosomes are chosen from the population based on their fitness. Thus, the better the fitness, the higher is the probability of a chromosome to be reproduced to the next generation. As a selection tool we apply the roulette wheel procedure discussed in the previous chapter.

### Crossover

Crossover is performed at a single point across the chromosome, selected using a uniform distribution. To preserve the validity of the chromosome, the crossover location is chosen in a quantity of three genes.

To avoid the possibility that even when the crossover operator is used the children chromosomes will be immediate replicas of their parents, we select $x$ from the set $\{1, \ldots, N_s - 1\}$. Thus, the first child solution consists of the first $3x \in C$ genes (columns) from the first parent and $N_c - 3x$ genes from the second parent, and vice versa for the second child.

We apply this operator with a probability $p_c$. Thus, the probability of two children chromosomes to be exact replicas of their parents is bounded from below by $1 - p_c$. This value is only a bound since there is a (low) probability that the exchanged genes between two parents will be identical.

### Mutation

The application of the mutation operator chosen in this study involves randomly changing one of the genes in the child chromosome. We mutate only the identity of one of the three vehicles participating in the simultaneous task. Thus, this operator is performed only on the first row of the chromosome. The identity of the new vehicle is selected randomly and we enforce the requirement that three different vehicles service each target. Different applications of this operator are possible, including the change of multiple genes in each reproduction stage. We apply this operator with a probability, denoted $p_m$, that is larger than zero. Note that a high value for $p_m$ will interfere with the crossover operator, as child solutions may not resemble the parent solutions that were chosen based on their fitness. In section 5.5 we examine the relationship between these two operators.

### Elitism

To avoid the possibility of losing the best solutions, when propagating to the new generation, we employ the elitism genetic operator and keep the best $N_e$ chromosomes from the previous generation. The rest of the new chromosomes $(N_f - N_e)$ are obtained by repeatedly producing children, by the methods described above, until the new population is filled.

## 5.5 CMTE Simulation Example

To clarify the CMTE scenario and illustrate the assignment algorithms presented in this chapter, a simulation example with a problem of size $V = 5$ and $N = 3$ is presented here. UAVs and targets were randomly placed in an area 110 km wide and 170 km long with an additional stand-off vehicle fixed directly north of the area of interest. Fig. 5.3(a) shows the targets and stand-in UAVs shortly after simulation start. Each target is shown with an associated detectability region (outlined in black) and cooperative tracking region (solid gray pie wedges). Recall that the cooperative tracking region is the intersection of the detectability region with the LOS angles greater than 45° different from the stand-off vehicle LOS. Task time availability windows are computed based on computing minimum

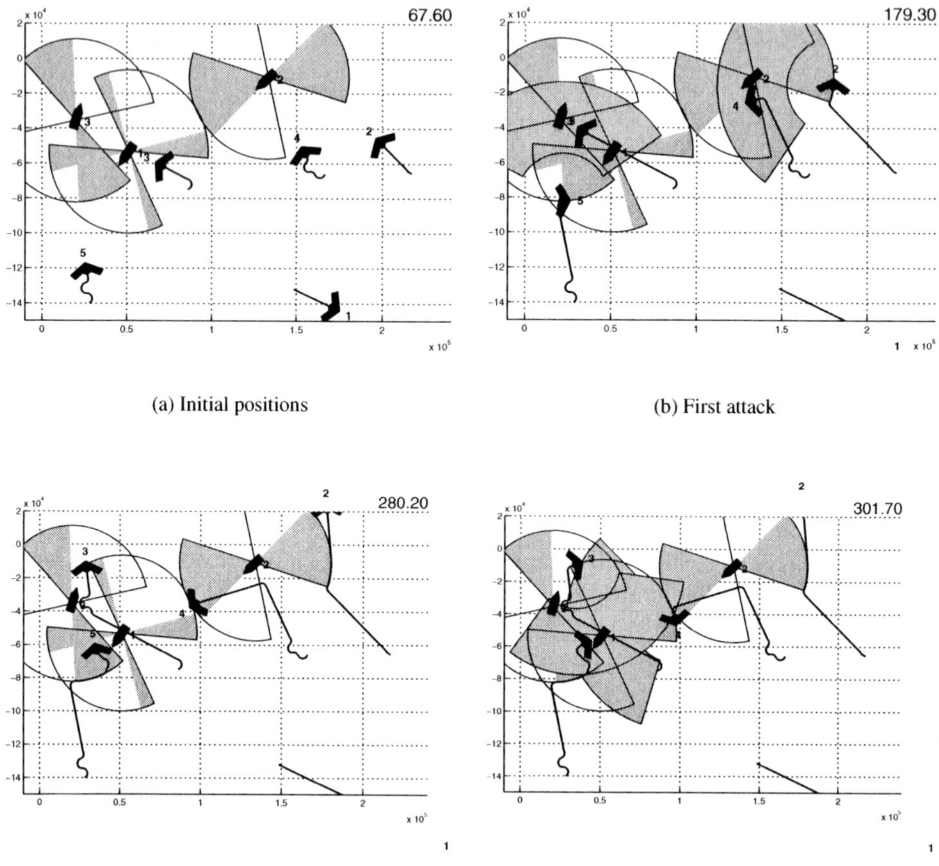


(a) Initial positions

(b) First attack

(c) Second attack

(d) Final attack

**Figure 5.3:** CMTE example with MILP task scheduling.

time trajectories to these regions. All target locations are known at the start of the simulation run, due to the ability of the stand-off vehicle to detect and track them. The mission plan illustrated in Fig. 5.3 was calculated using the MILP formulation presented in section 5.3, but a GA-generated example would result in similar or even identical behavior. This CMTE mission required 70 variables (67 binary, 3 continuous) and 81 constraints to solve the assignment and scheduling problem using a MILP formulation. [3]

The CMTE scenario is rich in timing complexity, making visualization difficult without animation. We therefore present several time-phased plots to illustrate the time-coordinated nature of the vehicles actions. Fig. 5.3 shows the progression of the simulated scenario at four distinct points in time. Fig. 5.3(a) shows the vehicle and target positions shortly after the simulation start. Vehicles 5 and 3 are assigned to prosecute Target 3. Vehicles 2 and 4 and assigned to prosecute Target 2. Vehicle 1 has no assignment. Vehicles 4 and 5 have added some loiter to their paths so that they can begin tracking at the same time as weapons are launched at their targets. The assignment to prosecute Target 1 has also been made, although it will not be prosecuted until the other two targets have been attacked, and vehicles are free to act on Target 1.

Fig. 5.3(b) shows the prosecution of Targets 2 and 3. Vehicles 5 and 2 are tracking the targets, in cooperation with the stand-off vehicle. Sensor footprint areas are shown as gray sections of a disk. Vehicle 4 has already launched a weapon at Target 2 and turned off toward its next task. Vehicle 3 is about to launch at Target 3. Vehicle 1 is not involved in the target prosecution.

Fig. 5.3(c) shows the vehicle trajectories up to $t = 280.2$ seconds. Targets 2 and 3 have been destroyed and are no longer being tracked. Vehicle 5 is loitering just outside of attack range on Target 1 while waiting for Vehicles 3 and 4 to move into position on the tracking circle for Target 1. Fig. 5.3(d) shows the prosecution of Target 3. Vehicle 5 has launched a weapon at Target 1, and Vehicles 3 and 4 are cooperatively tracking the target. The stand-off vehicle is not needed for this tracking, as two stand-in vehicles are engaged in tracking the target. Two stand-in vehicles are used as this enables the cooperative tracking, and thus the required attack, to start earlier than if the tracking was performed only by one of the stand-in vehicles, plus the stand-off vehicle. Due to the speed and heading of Target 1, the gray regions where a stand-in vehicle can track in cooperation with the stand-off vehicle are quite small.

## 5.6 Summary

The problem of assigning UAVs to simultaneously prosecute tasks has been presented. The problem consists of multiple ground moving targets prosecuted by a team of heterogeneous UAVs, carrying designated sensors, where all of the UAVs but one carry weapons. The problem was solved using MILP and GA. It was demonstrated that for scenarios involving a few vehicles and targets the problem can be solved in sufficient time using the MILP. For larger-sized problems the computational complexity renders MILP infeasible due to timing constraints on the simultaneous tasks and the coupling between task assignment and path planning for each UAV. For such problems a GA can be used.

## Bibliography

[1] M. Darrah, W. Niland, and B. Stolarik. Multiple UAV dynamic task allocation using mixed-integer linear programming in a SEAD mission. In *Proceedings of the AIAA Infotech Conference*, Alexandria, VA, 2005.

[2] L. E. Dubins. On curves of minimal length with a constraint on average curvature, and with prescribed initial and terminal position. *American Journal of Mathematics*, 79(3): 497–516, July 1957.

[3] D. B. Kingston and C. J. Schumacher. Time-dependent cooperative assignment. In *Proceedings of the American Control Conference*, Portland, OR, 2005.

[4] C. Schumacher, P. Chandler, S. Rasmussen, and D. Walker. Path elongation for UAV task assignment. In *Proceedings of the AIAA Guidance, Navigation, and Control Conference*, Austin, TX, 2003.

[5] C. J. Schumacher, S. J. Rasmussen, and D. H. Walker. Path elongation for UAV task assignment. In *Proceedings of the AIAA Guidance, Navigation, and Control Conference*, Austin, TX, 2003.

[6] T. Shima and C. Schumacher. Assigning cooperating UAVs to simultaneous tasks on consecutive targets using genetic algorithms. *Journal of the Operational Research Society*, 2008.

Chapter 6

# Estimation Algorithms for Improved Cooperation Under Uncertainty

*Tal Shima and Steven Rasmussen*

*It is not certain that everything is uncertain.*
—Blaise Pascal

Information flow imperfections, such as communication delays, may produce different information sets for the different UAVs in the group leading to multiple strategies. The result can be uncoordinated assignments, such as multiple vehicles wrongly assigned to perform the same task on a certain target, leaving other tasks unassigned. In this chapter, we present a decision-estimation architecture for a team of agents cooperating under uncertainties arising from imperfections in vehicle state measurements and intervehicle communications. For the estimation process we assume that the communication imperfections cause uncertainty on teammates' position and their respective updated costs to prosecute known targets. Thus, we concentrate on estimating the position of teammates. To motivate the use of the decision-estimation architecture, we describe its application using a redundant centralized implementation (RCI) of an iterative capacitated transshipment assignment problem (CTAP) algorithm. The next section describes the method of RCI, and more information on the iterative CTAP can be found in Chapter 3, section 3.3.

## 6.1 Redundant Centralized Implementation

Centralized assignment algorithms, such as the *iterative CTAP*, described in Chapter 3, section 3.3, can be implemented in an RCI manner. RCI replicates the centralized algorithm across all the vehicles and each one runs the algorithm. Assuming all the vehicles have the same information when they run the algorithm, all the members of the group will come up with the same plans, can implement their portion of the plans, and thus can act in a coordinated manner. For the iterative CTAP, the algorithm needs to be run when new targets are found or when a task has not been accomplished successfully. When such an event happens, each vehicle sends its current state, consisting of position, heading, and

**Table 6.1:** Cost matrix example at stage $5 \in S$ of the assignment.

| $V \setminus T$ | 1 | 2 | 3 | 4 |
|---|---|---|---|---|
| 1 | 11000 | 9000 | 1100 | 4000 |
| 2 | 1500 | 900 | 9200 | 3200 |
| 3 | 1000 | 1000 | 9000 | 5000 |

speed, to the rest of the group. Each vehicle uses the information gathered from all UAVs in the group to independently run the assignment algorithm.

## 6.2 Effect of Communication Delays

To obtain the group assignment using the iterative CTAP algorithm, a number of iterations, equal to the number of group tasks, are performed. During each iteration, the cost is calculated for each vehicle to perform a task on a target at the current stage

$$c_{l,i,j}^{X_{l-1}} = f(\boldsymbol{x}^i, \boldsymbol{t}^j) \qquad \forall i \in U,\, j \in T, \tag{6.1}$$

where $\boldsymbol{x}^i$ and $\boldsymbol{t}^j$ are the states of vehicle $i$ in the set of vehicles and target $j$ in the set of targets, respectively. The state of the target includes not only its position but also the task that needs to be performed on it based on the set of assignments that have already been defined $X_{l-1}$. The state of the vehicle includes its expected position and heading at the beginning of stage $l$ based on the set of assignments up to this stage ($X_{l-1}$). At each iteration a cost matrix is formed composed of all vehicle to target combinations, and the minimum element is sought, defining the vehicle to target assignment at this stage. Note that here we choose the minimum cost formulation, whereas in Chapter 3 we used the maximum benefit cost.

An example of such a cost matrix is plotted in Table 6.1 for a scenario containing three UAVs and four targets. In this example, the assignment algorithm is at the fifth iteration ($l = 5$), where classification and attack have already been assigned for Targets 2 and 4. Since it is the minimum cost in the cost matrix, the assignment selected at this stage is Vehicle 2 servicing Target 2.

For this example, the cost matrix at the next stage is plotted in Table 6.2. The second column is empty as all the tasks on target $2 \in T$ have been assigned (classification, attack, and verification). The chosen single assignment at this stage is of Vehicle 3 performing classification on Target 1.

Note that only the costs of the second row have been updated as that row relates to vehicle $2 \in V$, which performed the previous task. As explained above, in this iterative process, after a vehicle has been assigned to a task its position and heading at the end of the task are computed, serving as the initial conditions for computing the cost to perform the next task. All these calculations are performed by each vehicle in the group, running the iterative network flow algorithm, based only on the communicated information of each vehicle's state at the beginning of the assignment.

**Table 6.2:** Cost matrix example at stage $6 \in S$ of the assignment.

| $V \setminus T$ | 1 | 2 | 3 | 4 |
|---|---|---|---|---|
| 1 | 11000 | | 1100 | 4000 |
| 2 | 1900 | | 9800 | 3700 |
| 3 | 1000 | | 9000 | 5000 |

**Table 6.3:** Cost matrix with time stamp example; stage $5 \in S$.

| $V \setminus T$ | | 1 | 2 | 3 | 4 |
|---|---|---|---|---|---|
| 1 | 21 | 11000 | 9000 | 1100 | 4000 |
| 2 | 23 | 1500 | 900 | 9200 | 3200 |
| 3 | 21 | 1000 | 1000 | 9000 | 5000 |

For the UAVs to arrive with the same group plan, and thus act coordinately, a synchronized database of target and vehicle state information is required. As noted earlier, in this study we assume that replanning occurs only when a new target is found or when a task fails. Information regarding such an event is transmitted to the group by the relevant vehicle, triggering communication from all vehicles of their current state of position and heading.

In a realistic scenario, communication constraints, such as communication delays, are expected. An example of the cost matrix available to Vehicle 2 at stage $5 \in S$ is plotted in Table 6.3, where a cost time stamp is given in the second column. This is the time at which the costs have been calculated. Note that the second row, corresponding to tasks that can be performed by Vehicle 2, are computed based on its current information, while the other rows are based on delayed information. Assuming that the other rows (of Vehicles 1 and 3) have been computed at the same time; then it is possible for Vehicle 2 to synchronize the costs by computing its cost at that earlier time. This might lead to an assignment in this stage that is different than for Vehicle 2 to service Target 2.

Implementing this synchronization mechanism in the group will lead to UAV actions based on the same plan. However, any of the assigned trajectories based on old positions would be infeasible. This could invalidate the entire group coordination. Thus, in this study, for each vehicle to compute a flyable trajectory for itself, each vehicle uses its own current state and the delayed states received from teammates for the task assignment. Consequently, each vehicle acts based on a different information set, leading possibly to multiple strategies in the group. The result can be uncoordinated assignments, such as multiple vehicles wrongly assigned to perform the same task on a certain target, leaving other tasks unassigned [2].

In the following sections estimation algorithms, derived in [6], are presented. These algorithms enable better synchronization between the databases of the different vehicles and thus promote coordination.

## 6.3 Communication and Estimation Models

In the course of studying the effects of communication on cooperative control [3, 4, 5], a communications model was developed and implemented in the MultiUAV2 simulation.[1] This communications model was designed to produce the effects of realistic communications that are important to cooperative control systems, such as delayed and lost messages. The effects-based communication simulation used here is that found in [2], where messages are passed to simulate vehicle communication at each major model update.[2] A broadcast communication model is implicitly assumed for the vehicle communication. While not specifically targeted to address a particular physical implementation, such a model encompasses the typical view that the communications are time-division multiplexed.

As a consequence of the major model update, we define the *data rate*, at a given simulation step, as the total size of the messages collected, in bits, divided by the duration of the model update, yielding a rate in bits/s. Currently, all message data are represented in MATLAB using double-precision floating-point numbers, and in the computation of data rate, the message overhead is not considered, only the message payload. In a physical communication implementation there would be considerably more overhead, e.g., redundancy, error correction, or encryption. Thus, retaining double-precision in the ideal communication model remains a reasonable indicator of data rates, particularly since we are interested only in an initial estimate and, more important, a relative comparison of communication necessary for cooperative operations under various conditions.

We use the Dubins' car model, presented in section B.4, to represent the constant speed motion of each UAV.

We assume that each UAV has measurements only on its own position $(x, y)$ in a Cartesian inertial reference frame. The measurement equation is

$$z_k = \begin{bmatrix} x \\ y \end{bmatrix} + \boldsymbol{v}_k, \tag{6.2}$$

and $\{\boldsymbol{v}_k\}$, $\boldsymbol{v} \in \mathbb{R}^2$, is a zero mean white sequence with positive definite covariance matrix $R_k$. We denote the measurement rate as $f_1$.

For the estimation process we will use the state vector

$$\boldsymbol{x} \equiv \begin{bmatrix} x & y & v_x & v_y \end{bmatrix}^T \tag{6.3}$$

with the dynamics

$$\dot{\boldsymbol{x}} = \boldsymbol{A}\boldsymbol{x} + \boldsymbol{B}(\boldsymbol{x})u + \boldsymbol{G}\omega, \tag{6.4}$$

where

$$\boldsymbol{A} = \begin{bmatrix} 0 & 0 & 1 & 0 \\ 0 & 0 & 0 & 1 \\ 0 & 0 & 0 & 0 \\ 0 & 0 & 0 & 0 \end{bmatrix}, \quad \boldsymbol{B}(\boldsymbol{x}) = \begin{bmatrix} 0 \\ 0 \\ -\Omega_{max} v_y \\ \Omega_{max} v_x \end{bmatrix}, \quad \boldsymbol{G} = \begin{bmatrix} 0 & 0 \\ 0 & 0 \\ 1 & 0 \\ 0 & 1 \end{bmatrix}, \tag{6.5}$$

and $\Omega_{max}$ is the maximum turning rate of the vehicle. Note the noise on the velocity components as some uncertainty is expected, for example, due to wind.

[1] For more information on the communication model implementation see Appendix A, section A.3.4.

[2] The major model update in MultiUAV2 is 10 Hz.

We assume that each UAV knows the control actions of his teammates, since they are all acting based on the same cooperative team task assignment plan. Moreover, since all vehicles implement Dubins' optimal trajectories, these controls are restricted to the set

$$u \in \{-1, 0, 1\}, \tag{6.6}$$

as explained in Appendix B.

Note that in the above representation of the equations of motion the dynamics matrix $\boldsymbol{A}$ is constant while $\boldsymbol{B}$ is state dependent. Thus, $\boldsymbol{A}$ and $\boldsymbol{G}$, are identical to all vehicles, enabling the computationally efficient algorithm presented in the next section.

## 6.4 Efficient Cooperative Estimation

In this section we present communication and computation efficient algorithms for estimating the position and velocity of teammates. Having these estimates will allow continuous estimation of the costs for vehicles to service new targets based on prior communicated information. This will make it possible to immediately produce assignments when replanning is needed. The estimate will not be performed by a central agent. Rather, it will be performed, in a decentralized manner, by each member in the group.

### 6.4.1 Computationally Efficient Information Filter

In this subsection we assume that each UAV receives synchronous updates from teammates at a constant rate $f_2$, where $f_2 < f_1$. The transmitted or received data can be the raw measurements; better yet, the data can be the new information gained since the last transmission. Both will be discussed next.

A well-known technique in decentralized estimation is the information filter (IF) [1]. It is algebraically equivalent to the Kalman filter (KF) with a computationally simpler observation update stage at the cost of increased complexity in the time update stage.

The KF algorithm generates estimates for the state, denoted as $\hat{\boldsymbol{x}}$, together with a corresponding estimation covariance, denoted as $\boldsymbol{P}$. The IF, being a representation of the KF from the information viewpoint, uses the following definitions:

$$\boldsymbol{Y}_{(\cdot)/(\cdot)} \equiv \boldsymbol{P}^{-1}_{(\cdot)/(\cdot)}, \tag{6.7}$$

$$\hat{\boldsymbol{y}}_{(\cdot)/(\cdot)} \equiv \boldsymbol{Y}_{(\cdot)/(\cdot)}\hat{\boldsymbol{x}}_{(\cdot)/(\cdot)}, \tag{6.8}$$

$$\boldsymbol{I}_k \equiv \boldsymbol{H}_k^T \boldsymbol{R}_k^{-1} \boldsymbol{H}_k, \tag{6.9}$$

$$\boldsymbol{i}_k \equiv \boldsymbol{H}_k^T \boldsymbol{R}_k^{-1} z_k, \tag{6.10}$$

where $\boldsymbol{Y}$ is the well-known Fisher information matrix, $\hat{\boldsymbol{y}}$ is the new estimated state vector, $I$ is the a priori expected information held in each measurement, and $i$ is the a posteriori actual information held in a single measurement.

We will denote the estimation of the state vector of UAV $j \in U$ by UAV $i \in U$ as $\hat{\boldsymbol{x}}^{i,j}$, and correspondingly the IF state vector is denoted $\hat{\boldsymbol{y}}^{i,j}$. Note that $\hat{\boldsymbol{x}}^{i,i}$, represents the estimated state vector of UAV $i \in U$ by itself, based on the information transmitted to teammates. The estimation of one's own state, based on all available measurements, is denoted $\hat{\boldsymbol{x}}^i$. The distinction between $\hat{\boldsymbol{x}}^i$ and $\hat{\boldsymbol{x}}^{i,i}$ will become clear.

The equations of the $N_v$ IFs run by each UAV $i \in U$, based on the communicated information, at a rate of $f_2$, are

Time Update:

$$\hat{\boldsymbol{y}}^{i,j}_{k+1/k} = \left[\mathbf{1} - \boldsymbol{\Omega}^{i,i}_k \boldsymbol{\Gamma}^T_k\right] \boldsymbol{\Phi}^{-T}_k \left[\hat{\boldsymbol{y}}^{i,j}_{k/k} + \boldsymbol{Y}^{i,i}_{k/k} \boldsymbol{\Phi}^{-1}_k \boldsymbol{B}_k \left(\boldsymbol{Y}^{i,i}_{k/k}{}^{-1} \boldsymbol{y}^{i,j}_{k/k}\right) u^j_k\right] \quad \forall j \in U, \tag{6.11}$$

$$\boldsymbol{Y}^{i,i}_{k+1/k} = \boldsymbol{M}^{i,i}_k - \boldsymbol{\Omega}^{i,i}_k \boldsymbol{\Gamma}_k{}^T \boldsymbol{M}^{i,i}_k, \tag{6.12}$$

where

$$\boldsymbol{M}^{i,i}_k = \boldsymbol{\Phi}_k{}^{-T} \boldsymbol{Y}^{i,i}_{k/k} \boldsymbol{\Phi}_k{}^{-1}, \tag{6.13}$$

$$\boldsymbol{\Omega}^{i,i}_k = \boldsymbol{M}^{i,i}_k \boldsymbol{\Gamma}_k \left[\boldsymbol{\Gamma}^T_k \boldsymbol{M}^{i,i}_k \boldsymbol{\Gamma}_k + \boldsymbol{Q}^{-1}_k\right]^{-1}; \tag{6.14}$$

$\boldsymbol{\Phi}_k$, $\boldsymbol{B}_k$, and $\boldsymbol{\Gamma}_k$ are the discrete versions of $\boldsymbol{A}$, $\boldsymbol{B}$, and $\boldsymbol{G}$, respectively; and $u^j_k$ is the control action of UAV $j \in U$ known to UAV $i \in U$ (the one performing the estimation process), since it is assumed that all the UAVs abide by the same task assignment plan.

Observation Update:

$$\hat{\boldsymbol{y}}^{i,j}_{k+1/k+1} = \hat{\boldsymbol{y}}^{i,j}_{k+1/k} + \boldsymbol{i}^j_{k+1} \quad \forall j \in U, \tag{6.15}$$

$$\boldsymbol{Y}^{i,i}_{k+1/k+1} = \boldsymbol{Y}^{i,i}_{k+1/k} + \boldsymbol{I}^j_{k+1}. \tag{6.16}$$

Since the quality of information obtained from each UAV is assumed identical, then the information matrix is identical. Thus, Eqs. (6.12), (6.13), (6.14), and (6.16) are computed only once for all filters, reducing considerably the computational effort.

Each UAV also runs another IF on its own states using its measurements at a rate of $f_1 > f_2$. The equations for the state vector $\boldsymbol{y}^i$ and information matrix $\boldsymbol{Y}^i$ of this filter are

Time Update:

$$\hat{\boldsymbol{y}}^{i}_{k+1/k} = \left[\mathbf{1} - \boldsymbol{\Omega}^{i}_k \boldsymbol{\Gamma}^T_k\right] \boldsymbol{\Phi}^{-T}_k \left[\hat{\boldsymbol{y}}^{i}_{k/k} + \boldsymbol{Y}^{i}_{k/k} \boldsymbol{\Phi}^{-1}_k \boldsymbol{B}_k (\boldsymbol{Y}^{i}_{k/k}{}^{-1} \boldsymbol{y}^{i}_{k/k}) u^i_k\right], \tag{6.17}$$

$$\boldsymbol{Y}^{i}_{k+1/k} = \boldsymbol{M}^{i}_k - \boldsymbol{\Omega}^{i}_k \boldsymbol{\Gamma}^T_k \boldsymbol{M}^{i}_k, \tag{6.18}$$

where

$$\boldsymbol{M}^{i}_k = \boldsymbol{\Phi}_k^{-T} \boldsymbol{Y}^{i}_{k/k} \boldsymbol{\Phi}_k^{-1}, \tag{6.19}$$

$$\boldsymbol{\Omega}^{i}_k = \boldsymbol{M}^{i}_k \boldsymbol{\Gamma}_k \left[\boldsymbol{\Gamma}^T_k \boldsymbol{M}^{i}_k \boldsymbol{\Gamma}_k + \boldsymbol{Q}^{-1}_k\right]^{-1}; \tag{6.20}$$

Observation Update:

$$\hat{\boldsymbol{y}}^{i}_{k+1/k+1} = \hat{\boldsymbol{y}}^{i}_{k+1/k} + \boldsymbol{i}^j_{k+1} \quad \forall j \in U, \tag{6.21}$$

$$\boldsymbol{Y}^{i}_{k+1/k+1} = \boldsymbol{Y}^{i}_{k+1/k} + \boldsymbol{I}^j_{k+1}. \tag{6.22}$$

The transmitted information to the group is $\boldsymbol{i}^j_k$ and $\boldsymbol{I}^j_k$. If this information is the current measurement, then $\boldsymbol{I}^j_k$ and $\boldsymbol{i}^j_k$ can be computed based on Eqs. (6.9) and (6.10), respectively. However, it will be more beneficial to send all the information gathered since the last transmission. Such information can be computed as

$$\boldsymbol{I}_k^j = \boldsymbol{Y}_{k/k}^j - \boldsymbol{Y}_{k/k}^{j,j}, \tag{6.23}$$

$$\boldsymbol{i}_k^j = \boldsymbol{y}_{k/k}^j - \boldsymbol{y}_{k/k}^{j,j}. \tag{6.24}$$

Note that compared to a full-blown set of $N_u$ IFs on team members' states (running at the given update rate of $f_2$ and calculating Eqs. (6.11) through (6.16) $N_u$ times), this efficient algorithm induces the same communication load but with reduced computational load (due to the single calculation of Eqs. (6.12) through (6.14), and (6.16)). This reduction does not affect the estimation accuracy.

### 6.4.2 Communication Efficient Information Filter

Now, assume that each UAV receives asynchronous updates from teammates. These updates, $\boldsymbol{I}_k^j$ and $\boldsymbol{i}_k^j$, are based on the quality of the current state estimate of UAV $j \in U$ by UAV $i \in U$ as expected by UAV $j \in U$. Thus, the information is sent to the group members by UAV $j \in U$ only if

$$\boldsymbol{e}_j^T \boldsymbol{E} \boldsymbol{e}_j > \epsilon, \tag{6.25}$$

where

$$\boldsymbol{e}_j = \boldsymbol{Y}_{k/k}^{j}{}^{-1} \boldsymbol{y}_{k/k}^j - \boldsymbol{Y}_{k/k}^{j,j}{}^{-1} \boldsymbol{y}_{k/k}^{j,j}, \tag{6.26}$$

$$\boldsymbol{E} = \begin{bmatrix} 1 & 0 & 0 & 0 \\ 0 & 1 & 0 & 0 \\ 0 & 0 & 0 & 0 \\ 0 & 0 & 0 & 0 \end{bmatrix}, \tag{6.27}$$

$\epsilon$ is a design parameter, and $E$ was chosen so that the error will be defined between the position estimates.

The equations of the $N_v$ IFs, run by each UAV $i \in U$ based on the communicated information, are

Time Update:

$$\hat{\boldsymbol{y}}_{k+1/k}^{i,j} = \left[\boldsymbol{1} - \boldsymbol{\Omega}_k^{i,j} \boldsymbol{\Gamma}_k^T\right] \boldsymbol{\Phi}_k^{-T} \left[\hat{\boldsymbol{y}}_{k/k}^{i,j} + \boldsymbol{Y}_{k/k}^{i,j} \boldsymbol{\Phi}_k^{-1} \boldsymbol{B}_k \left(\boldsymbol{Y}_{k/k}^{i,j}{}^{-1} \boldsymbol{y}_{k/k}^{i,j}\right) u_k^j\right] \quad \forall j \in U, \tag{6.28}$$

$$\boldsymbol{Y}_{k+1/k}^{i,j} = \boldsymbol{M}_k^{i,j} - \boldsymbol{\Omega}_k^{i,j} \boldsymbol{\Gamma}_k^T \boldsymbol{M}_k^{i,j} \quad \forall j \in U, \tag{6.29}$$

where

$$\boldsymbol{M}_k^{i,j} = \boldsymbol{\Phi}_k^{-T} \boldsymbol{Y}_{k/k}^{i,j} \boldsymbol{\Phi}_k^{-1} \quad \forall j \in U, \tag{6.30}$$

$$\boldsymbol{\Omega}_k^{i,j} = \boldsymbol{M}_k^{i,j} \boldsymbol{\Gamma}_k \left[\boldsymbol{\Gamma}_k^T \boldsymbol{M}_k^{i,j} \boldsymbol{\Gamma}_k + \boldsymbol{Q}_k^{-1}\right]^{-1} \quad \forall j \in U; \tag{6.31}$$

Observation Update:

$$\hat{\boldsymbol{y}}_{k+1/k+1}^{i,j} = \hat{\boldsymbol{y}}_{k+1/k}^{i,j} + \boldsymbol{i}_{k+1}^j \quad \forall j \in U, \tag{6.32}$$

$$\boldsymbol{Y}_{k+1/k+1}^{i,j} = \boldsymbol{Y}_{k+1/k}^{i,j} + \boldsymbol{I}_{k+1}^j \quad \forall j \in U. \tag{6.33}$$

**Table 6.4:** Simulation parameters.

| Estimation | Scenario |
|---|---|
| $f_1 = 10$ Hz | speed = 300ft/s |
| $f_2 = 0.2$ Hz | $\Omega_{max} = 0.15$ rad/sec |
| $\epsilon = 500$ ft | $Area = 20$ miles$^2$ |

### 6.4.3 Algorithm Implementation

The computation and communication efficient estimation algorithms are used to better synchronize the databases of the different vehicles in order to devise a coordinated group plan. Whenever replanning is needed, each vehicle $p \in U$ computes the estimated cost for its teammates to prosecute each target,

$$\hat{c}_{l,i,j}^{X_{l-1}} = f(\hat{\boldsymbol{x}}^{p,i}, \boldsymbol{t}^j) \qquad \forall i \in U \setminus \{p\}, \quad j \in T, \tag{6.34}$$

where $\hat{\boldsymbol{x}}^{p,i}$ is the estimated state vector of vehicle $i \in U \setminus \{p\}$ by vehicle $p \in U$ and $\boldsymbol{t}^j$ is the state of the assumed stationary target.

The vehicle also computes its own cost to prosecute all the targets based on its best estimate,

$$\hat{c}_{l,p,j}^{X_{l-1}} = f(\hat{\boldsymbol{x}}^p, \boldsymbol{t}^j) \qquad \forall j \in T. \tag{6.35}$$

It should be noted that, compared to the computation efficient algorithm, the communication efficient algorithm has a significantly larger computational complexity. This complexity results from the need for calculating Eqs. (6.29), (6.30), (6.31), and (6.33) $N_u$ times compared to only once for the computation efficient algorithm (see Eqs. (6.12), (6.13), (6.14), and (6.16)).

Compared to a full-blown set of IFs (composed of $N_u$ filters on team members' states), running with an observation update rate of $f_2$ the communication efficient algorithm has almost the same computational load (the only difference is the number of times simple equations (6.32) and (6.33) are computed) but much reduced communication load.

## 6.5 Simulation Study

The performance of the cooperative decision-estimation methodology was investigated through simulation. The scenario included three vehicles and six targets, and a pure communication delay of $t_d$ was implemented. The parameters used for running the different cases are given in Table 6.4. Next, some results are given. The detailed results can be found in [6].

First, a sample run is presented. The corresponding routes in the sample assignment are plotted in Fig. 6.1. The three vehicles start searching the terrain from left to right in a lawnmower fashion, where a square represents a target. The cooperative assignment is quite complicated and highly coupled. Following are the assignments for each vehicle to ensure the precedence requirements are not violated: Vehicle 1 classifies Target 4, classifies and

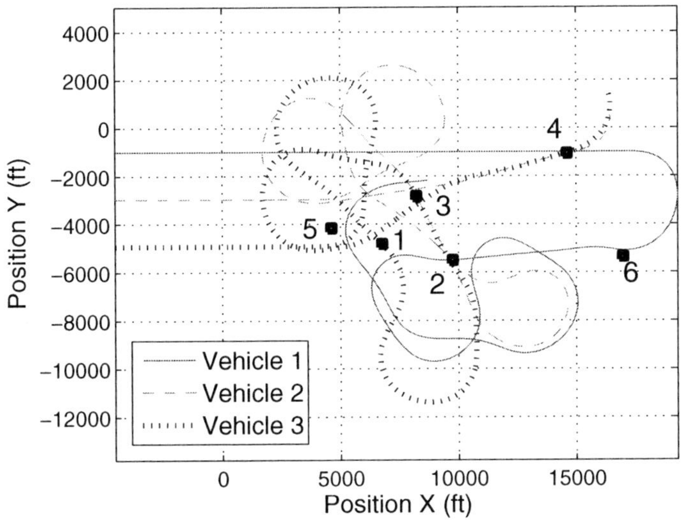


**Figure 6.1:** Example trajectories.

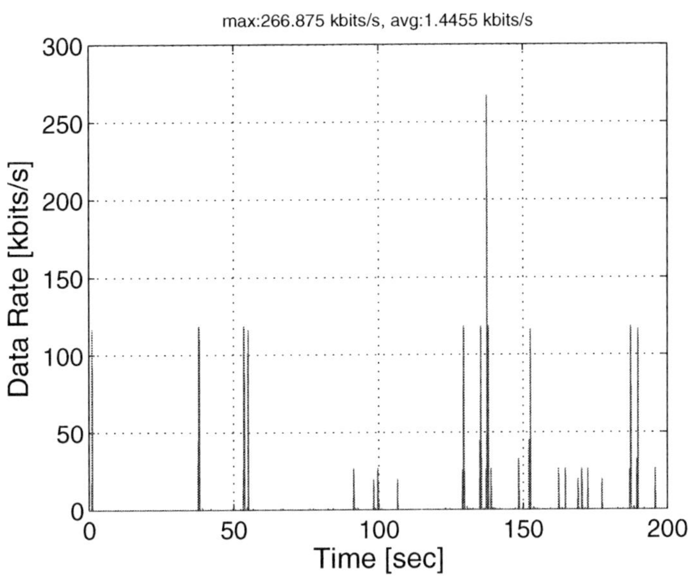


**Figure 6.2:** Communication sample run—no delay.

attacks Target 6, and attacks Target 2; Vehicle 2 verifies Target 5, classifies Target 2, verifies Target 6, classifies and attacks Target 3, and verifies Targets 1 and 4; Vehicle 3 classifies and attacks Targets 5 and 1, verifies Targets 2 and 3, and attacks Target 4.

In Figs. 6.2 through 6.4 the group communication data rate in the sample scenario is presented. Fig. 6.2 presents the communication load when there is no delay and there is no

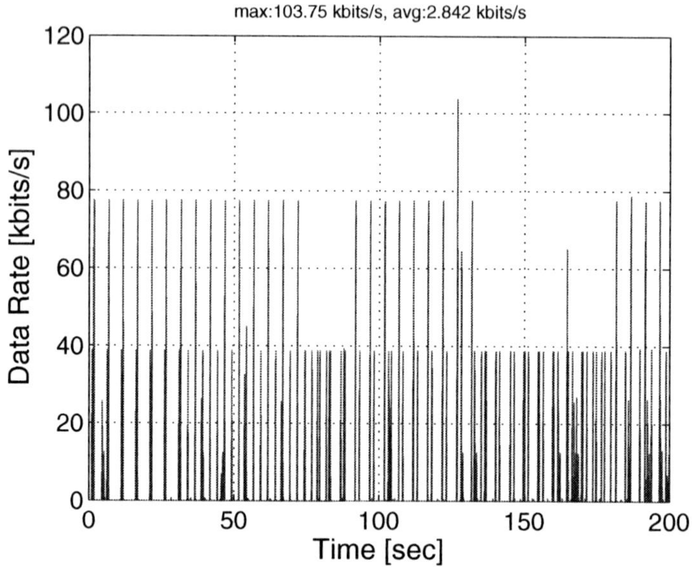


**Figure 6.3:** Communication sample run—computation efficient algorithm.

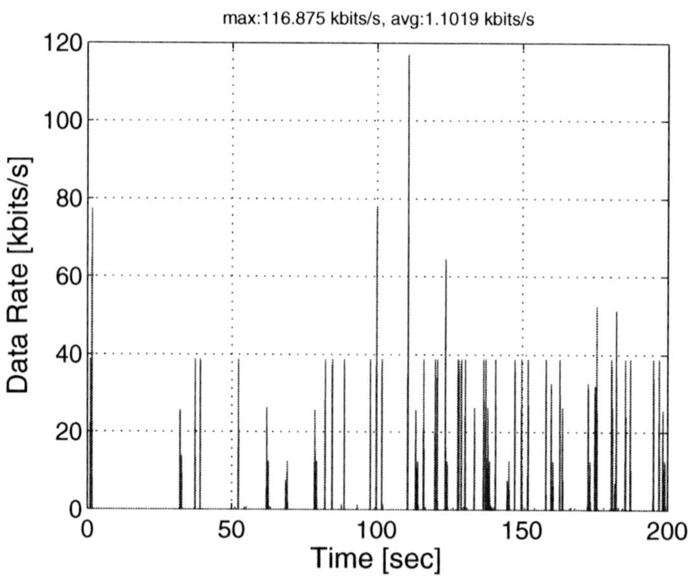


**Figure 6.4:** Communication sample run—communication efficient algorithm.

use of the estimation mechanisms proposed in section 6.4. The peaks of about 120 Kbits/sec occur each time a new target is found or when a task fails. In such an event all the vehicles need to communicate information regarding their location in order to reach a cooperative

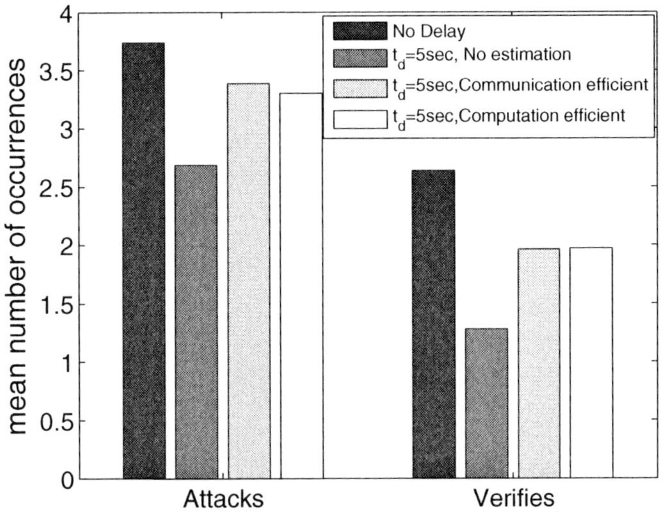


**Figure 6.5:** Average number of targets prosecuted.

plan. The maximum data rate in this scenario was 267 Kbits/sec and the average amount of communication was about 1.45 Kbits/sec. Similar qualitative results were obtained when a delay of 5 seconds was introduced ($t_d = 5$ sec) and no estimation mechanism was utilized. The results for the same scenario, but when the computation efficient algorithm is used, are plotted in Fig. 6.3. Notice that the peaks are lower (approximately 80 Kbits/sec) as communication regarding the vehicles' position is continuous and when a new target is found only that new information is broadcast. The maximum data rate in this run was 104 Kbits/sec and the average was 2.8 Kbits/sec. Fig. 6.4 represents the case when the communication efficient algorithm is used. Most of the peaks are about 40 Kbits/sec with the maximum approximately 117 Kbits/sec and the average 1.1 Kbits/sec. It is apparent that the amount of communication in this case is the lowest. Note that we used a given threshold ($\epsilon = 500$ ft) and that tuning of this parameter can be performed. This makes it possible to trade off between the amount of communication and group performance.

A Monte Carlo study, consisting of 200 runs, was used to evaluate the proposed decision-estimation methodology under communication delays. The random variables in these runs were the location of the targets and their appearance rate, and the initial location and heading of the UAV team members.

We compare the results of using the two efficient estimation algorithms to the case with no estimation mechanism. There is no need to compare the case when a full-blown set of $N_u$ IFs is used to estimate team members' states at a given update rate of $f_2$, as it has almost the same computation load as the communication efficient algorithm and the same communication load as the computation efficient algorithm. The only difference between the full-blown set of $N_u$ IFs and the communication efficient algorithm is the number of times simple equations (6.32) and (6.33) are calculated.

In Fig. 6.5 the performance of the group is evaluated by counting the average number of occurrences of target attacks and verifications in the allotted prosecution time of 200 seconds. The selected scenario time was 200 seconds for which, in most runs, only a part of the group mission was accomplished. The results for the classification task are not plotted—they are almost identical to those of attack. This is because in most cases, when a vehicle is assigned to classify a target it is also assigned next to attack it. The results are plotted for four cases: (i) perfect information (no delay), (ii) $t_d = 5$ sec and no estimation mechanism is employed, (iii) $t_d = 5$ sec and communication efficient algorithm employed, and (iv) $t_d = 5$ sec and computation efficient algorithm employed. Obviously the best performance, where approximately on average 3.8 targets are attacked and 2.6 targets are fully prosecuted (i.e., verified), is obtained for the case without delay. When there is a delay but no estimation algorithm is used, the performance deteriorates and only on average about 2.7 targets are attacked and 1.3 are fully prosecuted. As expected, using the proposed efficient estimation algorithms improves the performance and on average about 3.3 targets are attacked and 2 are fully prosecuted.

## 6.6 Summary

A decision-estimation methodology for a team of unmanned aerial vehicles cooperating under communication imperfections has been presented. It was shown that running filters on the states of teammates allows estimating their cost to prosecute new tasks. This enables better synchronization of the databases between the vehicles in the group and promotes coordination. For the estimation process two algorithms are presented, one computationally efficient and the other communication efficient. Performing a Monte Carlo study, using the MultiUAV2 test bed, the benefit of using the new algorithms under communication constraints was shown. They allow prosecution of more targets. Moreover, they enable controlling the communication bandwidth required for cooperation.

# Bibliography


[1] S. P. Maybeck. *Stochastic Models, Estimation, and Control, Vol.* 1, *Mathematics in Science and Engineering* 141, Academic Press, New York, 1979, 238–241.

[2] J. W. Mitchell and A. G. Sparks. Communication issues in the cooperative control of unmanned aerial vehicles. In *Proceedings of the* 41*st Annual Allerton Conference on Communication, Control, & Computing*, 2003.

[3] J. W. Mitchell, S. J. Rasmussen, and A. G. Sparks. Communication requirements in the cooperative control of wide area search munitions using iterative network flow. In *Theory on Algorithms for Cooperative Systems, Series on Computers and Operations Research*, Vol. 4, D. Grundel, R. Murphy, and P. Pardalos, eds., World Scientific, Singpore, 2004.

[4] J. W. Mitchell, S. J. Rasmussen, P. R. Chandler, and J. D. Redding. Synchronous and asynchronous communication effects on the cooperative control of uninhabited aerial vehicles. In *Proceedings of the AIAA Guidance, Navigation, and Control Conference*, Providence, RI, 2004.

[5] J. W. Mitchell, S. J. Rasmussen, and A. G. Sparks. Effects of target arrival rate on mission performance of cooperatively controlled UAVs with communication constraints. In *Proceedings of* 43*rd IEEE Conference on Decision and Control*, Paradise Island, Bahamas, 2004.

[6] T. Shima, S. Rasmussen, and P. Chandler. UAV team decision and control using efficient collaborative estimation. *ASME Journal of Dynamic Systems, Measurement, and Control*, 129(5):609–619, 2007.

## Chapter 7

# Effectiveness Measures for Operations in Uncertain Environments

***Brian Kish, Meir Pachter, and David Jacques***

*God does not play dice.*
—Albert Einstein

According to the United States Air Force's basic doctrine [1],

> *War is a complex and chaotic human endeavor. Uncertainty and unpredictability—what many call the "fog" of war—combine with danger, physical stress, and human fallibility to produce "friction", a phenomenon that makes apparently simple operations unexpectedly, and sometimes even insurmountably, difficult. Uncertainty, unpredictability, and unreliability are always present, but sound doctrine, leadership, organization, core personal values, technology and training can lessen their effects.*

Networked autonomous tactical UAVs address the "friction" elements of danger, physical stress, and human fallibility. The previous chapter addressed the fog element resulting from missing information. In this chapter, the fog element resulting from uncertain target and threat environments is addressed. As mentioned in Chapter 2, some cooperative team problems can be dominated by uncertainty. The challenge is to calculate the expected future value of a decision or action taken now. At the same time, actions now might decrease the level of uncertainty. Effectiveness measures, like those derived in this chapter, provide a foundation for the rational evaluation of cooperative rules of engagement or doctrine. Predictions based on a priori estimates are possible based on battle space models. Applied probability and optimal control theory are used to formulate and solve problems that maximize the probability of attacking at least $\hat{n}$ targets while constraining the probability of attacking at least $\hat{m}$ false targets. In addition to a value for effectiveness, the solutions yield optimal schedules for the system's operating point in terms of sensor threshold. So, not only will the benefit of assigning an agent to a task be known, but the operating point of

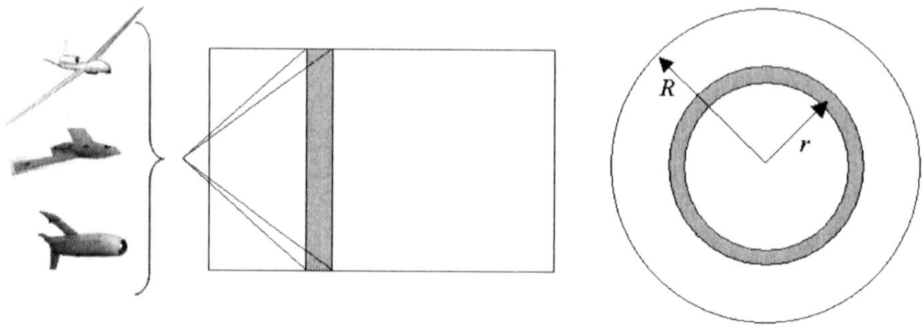


(a) Linear (b) Circular

**Figure 7.1:** Search patterns.

the agent will also be known. A variety of multiple target/multiple false target scenarios are considered, with varying assumptions on the probability distributions of the target and false targets. The generalized mathematical framework allows for search-and-destroy vehicles with $k$ warheads, where $k \in [1, \infty]$. Thus the effectiveness of munitions, tactical UAVs, and general sensor craft can all be evaluated.

## 7.1 Scenario

The search-and-destroy mission phases are searching, detecting, classifying, and attacking. Searching requires both a search pattern and area coverage rate. Two search patterns are used that handle a wide variety of battle space environments. The first pattern is the linear pattern depicted in Fig. 7.1(a). Although Fig. 7.1(a) shows a constant swath width, the linear pattern can be used for any linear-symmetric battle space. Further, any randomly shaped battle space can be broken up into piecewise linear-symmetric segments. The second pattern involves using concentric annuli emanating from the origin. Fig. 7.1(b) shows one such annulus at radius $r$. Searching a disc of radius $R$ using concentric annuli emanating from the origin is mathematically tractable and approximates an outward spiral search pattern. For constant-altitude flight, area coverage rate is simply the product of velocity and swath width. Area coverage rate is assumed specified by the designer or mission planner.

Detections occur based on assumed target and false target probability distributions. Uniform, Poisson, and circular-normal probability distributions are used. These distributions cover a wide variety of battle space environments where intelligence information can be vague or precise. Seven battle space scenarios are considered. Scenarios 1–4 use linear-symmetric battle spaces, and Scenarios 5–7 use circular battle spaces. The differences in the scenarios come from the target and false target probability distributions as well as the number of targets. For all scenarios, both targets and false targets are assumed fixed at unknown locations.

**Scenario 1** consists of a single target uniformly distributed in a linear-symmetric battle space among a Poisson field of false targets.

**Scenario 2** consists of a linear-symmetric battle space with a Poisson field of targets and a Poisson field of false targets.

**Table 7.1:** Scenario matrix.

| | Uniform Targets | Poisson Targets | Normal Targets |
|---|---|---|---|
| Uniform False Targets | Scenario 4 | — | — |
| Poisson False Targets | Scenarios 1, 3 | Scenario 2 | Scenarios 5, 7 |
| Normal False Targets | — | — | Scenario 6 |

**Scenario 3** consists of $N$ targets uniformly distributed in a linear-symmetric battle space among a Poisson field of false targets. *Note*: Scenario 1 is a special case of Scenario 3 with $N = 1$.

**Scenario 4** consists of $N$ targets and $M$ false targets uniformly distributed in a linear-symmetric battle space.

**Scenario 5** consists of $N$ targets normally distributed in a circular battle space among a Poisson field of false targets. Normally distributed refers to a circular-normal distribution centered at the origin with a target standard deviation $\sigma_y$.

**Scenario 6** consists of $N$ targets and $M$ false targets normally distributed in a circular battle space. Normally distributed refers to circular-normal distributions centered at the origin with a target standard deviation $\sigma_y$ and a false target standard deviation $\sigma_x$.

**Scenario 7** consists of a single target normally distributed in a circular battle space among a Poisson field of false targets. *Note*: Scenario 7 is a special case of Scenario 5 with $N = 1$.

Table 7.1 shows where the seven scenarios fall in a combination matrix for uniform, Poisson, and normal distributions. The techniques presented could easily be applied to any other combination.

## 7.2 Sensor Performance Modeling

When a sensor detects an object, it compares the image to a stored template or pattern and declares the object either a target or a false target. In practice, detected objects are classified to a certain level of discrimination. For example, one may classify an object as either air breathing or ballistic. A finer level of discrimination may be a specific type of air-breathing or ballistic object. Regardless of the level of discrimination, there is a point at which the sensor reports a detected object as either a target, thereby authorizing an attack, or a false target, thereby commanding no attack. Sensor performance is judged by how often the sensor is correct. The probability of a target report, $P_{TR}$, is the probability the sensor correctly reports a target when a target is present. The probability of a false target report, $P_{FTR}$, is the probability the sensor correctly reports a false target when a false target is present. Together, $P_{TR}$ and $P_{FTR}$ determine the entries of the binary "confusion matrix" shown in Table 7.2, which can be used to determine the outcome of a random draw each time an object is encountered in simulation.

The expression $(1 - P_{TR})$ represents the probability the sensor reports a false target when a target is present. This type of error results in a target not being attacked. The expression $(1 - P_{FTR})$ represents the probability the sensor reports a target when a false

**Table 7.2:** Simple binary confusion matrix.

| | Encountered Object | |
|---|---|---|
| | Target | False Target |
| Declared Object | | |
| Target | $P_{TR}$ | $1 - P_{FTR}$ |
| False Target | $1 - P_{TR}$ | $P_{FTR}$ |

target is present. This type of error results in a false target being attacked. For this binary confusion matrix, true positive fraction is $P_{TR}$, and false positive fraction is $(1 - P_{FTR})$. If multiple types of targets are involved, the dimension of the confusion matrix can be higher. This analysis only considers the $2 \times 2$ confusion matrix in Table 7.2.

The parameters $P_{TR}$ and $P_{FTR}$ are not independent. They are often linked by a receiver operating characteristic (ROC) curve. A mathematical representation of the ROC curve produces a graph of true positive fraction ($P_{TR}$) versus false positive fraction $(1 - P_{FTR})$ that starts at $(0, 0)$, then monotonically increases to $(1, 1)$. The following ROC curve model is used:

$$(1 - P_{FTR}) = \frac{P_{TR}}{(1 - c)\, P_{TR} + c}, \tag{7.1}$$

where the parameter, $c \in [1, \infty)$, will depend on the sensor and data processing algorithm. It will also depend on the vehicle speed (dwell time), and engagement geometry, which includes flight altitude and look angle. A family of ROC curves parameterized by $c$ is shown in Fig. 7.2. As $c$ increases, the ROC improves. As $c \to \infty$, the area under the curve approaches unity indicating perfect classification.

For a given sensor and algorithm with a constant area coverage rate, the operating point on a ROC curve is determined by the sensor's threshold for declaring a target. Although typically it must be determined from experimentation, it will be assumed the relation between $P_{TR}$ and $(1 - P_{FTR})$ is known and can be approximated using Eq. (7.1). Thus, controlling $P_{TR}$ is the goal.

## 7.3 Effectiveness Measures

The sequential events method produces a probability density function that can be integrated to calculate an overall effectiveness or expected level of collateral damage. To model the search-and-destroy mission, the following area definitions, illustrated in Fig. 7.3, are needed:

$$\begin{aligned} A_B &\equiv \text{battle space}, \\ A &\equiv \text{area covered through time } t \text{ or radius } r, \\ A_s &\equiv \text{area covered through time } T \text{ or radius } R, \\ dA &\equiv \text{area of sensor footprint at time } t \text{ or radius } r, \\ A_f &\equiv A_s - (A + dA). \end{aligned}$$

Choosing to define the sensor footprint as the infinitesimal area $dA$ allows for a physical interpretation of the mathematical operation of integration. One can think of

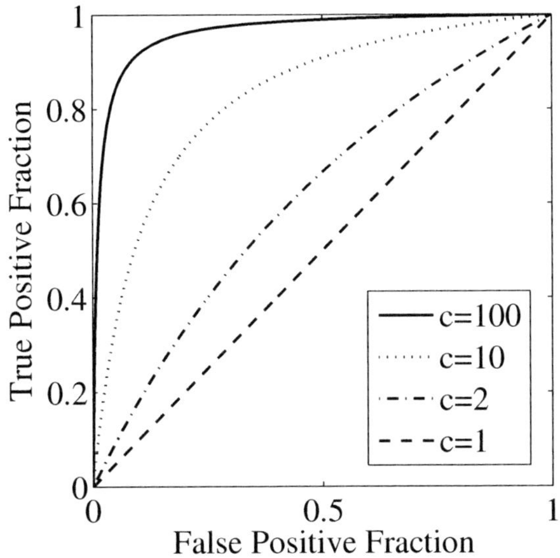


**Figure 7.2:** Family of ROC curves.

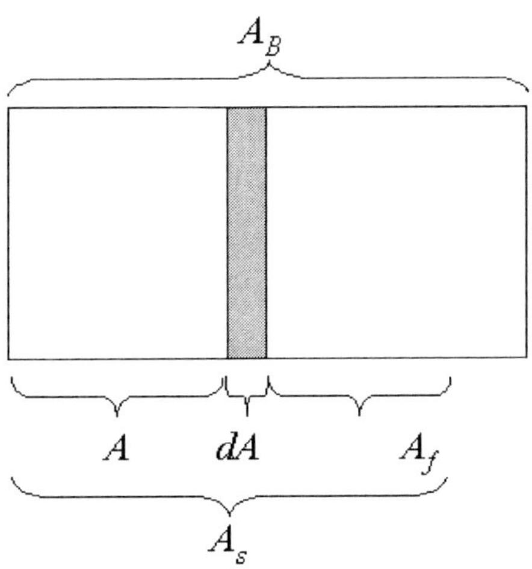


**Figure 7.3:** Area definitions.

dragging the sensor footprint over an area as an ordered search effort. For an event of interest (e.g., attacking a target) to occur in the sensor footprint at time $t$, certain events must occur prior to the event and certain events must occur after the event. For example, if all warheads were used prior to time $t$, there is no way an attack can occur at time $t$.

Therefore, events must be sequenced based on conditions. The various events of interest and associated numbering scheme are defined below.

$$\begin{aligned}
i &\equiv \text{number of false target encounters in } A,\\
j &\equiv \text{number of target encounters in } A,\\
k &\equiv \text{number of warheads available at the beginning of a mission},\\
m &\equiv \text{number of false target attacks},\\
M &\equiv \text{number of false targets in } A_B,\\
n &\equiv \text{number of target attacks},\\
N &\equiv \text{number of targets in } A_B,\\
\mathcal{M} &\equiv \text{events involving false target encounters},\\
\mathcal{N} &\equiv \text{events involving target encounters},\\
\mathcal{T} &\equiv \text{events where targets are correctly classified and attacked},\\
\mathcal{F} &\equiv \text{events where false targets are misclassified and attacked},\\
\overline{\mathcal{T}} &\equiv \text{events where targets are misclassified and not attacked},\\
\overline{\mathcal{F}} &\equiv \text{events where false targets are correctly classified and not attacked}.
\end{aligned}$$

For notation, the number of events occurring in a given area is subscripted along with the applicable area. For example, $n-1$ target attacks in $A$ is denoted $\mathcal{T}_{n-1,A}$. To calculate the probability of at least a certain number of attacks, one needs to know the probability of an exact number of attacks. One must distinguish between the case where all warheads are used and the case where warheads are left over. The probability of exactly $m$ false target attacks and $n$ target attacks when $m+n=k$ is denoted $P_{m,n}^{(m+n=k)}$. The probability of exactly $m$ false target attacks and $n$ target attacks when $m+n<k$ is denoted $P_{m,n}^{(m+n<k)}$. Both cases are needed to calculate the overall probability of an exact number of target and false target attacks in $A_s$. In general terms, for distributions involving $M$ false targets and $N$ targets, Decker [4] showed

$$\begin{aligned}
P_{m,n}^{(m+n=k)}(A_s) = \int_{A_s} \Big[ & P\left(\mathcal{T}_{n-1,A} \cap \overline{\mathcal{T}}_{j-(n-1),A} \cap \mathcal{N}_{N-1-j,A_f} \cap \mathcal{T}_{1,dA}\right) \\
& \times P\left(\mathcal{F}_{m,A} \cap \overline{\mathcal{F}}_{i-m,A} \cap \mathcal{M}_{M-i,A_f}\right) \\
& + P\left(\mathcal{T}_{n,A} \cap \overline{\mathcal{T}}_{j-n,A} \cap \mathcal{N}_{N-j,A_f}\right) \\
& \times P\left(\mathcal{F}_{m-1,A} \cap \overline{\mathcal{F}}_{i-(m-1),A} \cap \mathcal{M}_{M-1-i,A_f} \cap \mathcal{F}_{1,dA}\right)\Big]
\end{aligned} \tag{7.2}$$

and

$$\begin{aligned}
P_{m,n}^{(m+n<k)}(A_s) = \int_{A_s} \Big[ & P\left(\mathcal{T}_{n-1,A} \cap \overline{\mathcal{T}}_{j-(n-1),A} \cap \overline{\mathcal{T}}_{N-1-j,A_f} \cap \mathcal{T}_{1,dA}\right) \\
& \times P\left(\mathcal{F}_{m,A} \cap \overline{\mathcal{F}}_{i-m,A} \cap \overline{\mathcal{F}}_{M-i,A_f}\right) \\
& + P\left(\mathcal{T}_{n,A} \cap \overline{\mathcal{T}}_{j-n,A} \cap \overline{\mathcal{T}}_{N-j,A_f}\right) \\
& \times P\left(\mathcal{F}_{m-1,A} \cap \overline{\mathcal{F}}_{i-(m-1),A} \cap \overline{\mathcal{F}}_{M-1-i,A_f} \cap \mathcal{F}_{1,dA}\right)\Big],
\end{aligned} \tag{7.3}$$

where $i$ is the number of false target encounters in $A$ and $j$ is the number of target encounters in $A$[4]. The subtle difference between Eqs. (7.2) and (7.3) involves what must happen after time $t$ or radius $r$ in $A_f$. When $m+n=k$, once the final warhead is expended, one doesn't care if the vehicle comes across any more attack situations. An attack situation is one in which an attack would occur if the vehicle had a limitless supply of warheads. Therefore, one only cares about the number of remaining encounters in $A_f$. When $m+n<k$, there can be no attacks in $A_f$. Therefore, all subsequent target encounters must be misclassified, and all subsequent false target encounters must be correctly classified. In other words, $\mathcal{M}$'s and $\mathcal{N}$'s in Eq. (7.2) become $\overline{\mathcal{F}}$'s and $\overline{\mathcal{T}}$'s, respectively, in Eq. (7.3).

Calculating the probability of at least a certain number of attacks involves summing all the possible mutually exclusive probabilities. Thus,

$$P\left(m \geq \hat{m}\right) = \sum_{m=\hat{m}}^{\min(M,k-1)} \left[\sum_{n=0}^{\min(N,k-m-1)} P_{m,n}^{(m+n<k)}\right] + \sum_{m=\max\left(\hat{m},k-N\right)}^{k} P_{m,k-m}^{(m+n=k)} \tag{7.4}$$

and

$$P\left(n \geq \hat{n}\right) = \sum_{n=\hat{n}}^{\min(N,k-1)} \left[\sum_{m=0}^{\min(M,k-n-1)} P_{m,n}^{(m+n<k)}\right] + \sum_{n=\max\left(\hat{n},k-M\right)}^{k} P_{k-n,m}^{(m+n=k)}. \tag{7.5}$$

The cases in which either $m=0$ or $n=0$ must be considered separately. Either case results in a negative subscript in Eqs. (7.2) and (7.3). If a subscript is negative, the corresponding probability is simply zero. When all warheads are used,

$$\begin{aligned} P_{m=0,n=k}^{(m+n=k)}(A_s) = \int_{A_s} & \left[P\left(\mathcal{T}_{k-1,A} \cap \overline{\mathcal{T}}_{j-(k-1),A} \cap \mathcal{N}_{N-1-j,A_f} \cap \mathcal{T}_{1,dA}\right)\right. \\ & \left.\times P\left(\overline{\mathcal{F}}_{i,A} \cap \mathcal{M}_{M-i,A_f}\right)\right] \end{aligned} \tag{7.6}$$

and

$$\begin{aligned} P_{m=k,n=0}^{(m+n=k)}(A_s) = \int_{A_s} & \left[P\left(\mathcal{F}_{k-1,A} \cap \overline{\mathcal{F}}_{i-(k-1),A} \cap \mathcal{M}_{M-1-i,A_f} \cap \mathcal{F}_{1,dA}\right)\right. \\ & \left.\times P\left(\overline{\mathcal{T}}_{j,A} \cap \mathcal{N}_{N-j,A_f}\right)\right]. \end{aligned} \tag{7.7}$$

When warheads are left over,

$$P_{m=0,n<k}^{(m+n<k)}(A_s) = P\left(\mathcal{F}_{0,A_s}\right) \int_{A_s} P\left(\mathcal{T}_{n-1,A} \cap \overline{\mathcal{T}}_{j-(n-1),A} \cap \overline{\mathcal{T}}_{N-1-j,A_f} \cap \mathcal{T}_{1,dA}\right) \tag{7.8}$$

and

$$P_{m<k,n=0}^{(m+n<k)}(A_s) = P\left(\mathcal{T}_{0,A_s}\right) \int_{A_s} P\left(\mathcal{F}_{m-1,A} \cap \overline{\mathcal{F}}_{i-(m-1),A} \cap \overline{\mathcal{F}}_{M-1-i,A_f} \cap \mathcal{F}_{1,dA}\right). \tag{7.9}$$

An equivalent way to calculate $P\left(m \geq \hat{m}\right)$ is using

$$\begin{aligned} P\left(m \geq \hat{m}\right) &= 1 - P\left(m < \hat{m}\right) \\ &= 1 - \sum_{m=0}^{\min\left(M,\hat{m}-1\right)} \left[\sum_{n=0}^{\min(N,k-m-1)} P_{m,n}^{(m+n<k)}\right] - \sum_{m=\max(0,k-N)}^{\min\left(M,\hat{m}-1\right)} P_{m,k-m}^{(m+n=k)}, \end{aligned} \tag{7.10}$$

**Table 7.3:** Twelve elemental probabilities.

| Name | Elemental Probability | Condition |
|---|---|---|
| $P_1$ | $P\left(\mathcal{T}_{n-1,A} \cap \overline{\mathcal{T}}_{j-(n-1),A} \cap \mathcal{N}_{N-1-j,A_f} \cap \mathcal{T}_{1,dA}\right)$ | $P_{m,n}^{(m+n=k)}$ |
| $P_2$ | $P\left(\mathcal{F}_{m,A} \cap \overline{\mathcal{F}}_{i-m,A} \cap \mathcal{M}_{M-i,A_f}\right)$ | $P_{m,n}^{(m+n=k)}$ |
| $P_3$ | $P\left(\mathcal{T}_{n,A} \cap \overline{\mathcal{T}}_{j-n,A} \cap \mathcal{N}_{N-j,A_f}\right)$ | $P_{m,n}^{(m+n=k)}$ |
| $P_4$ | $P\left(\mathcal{F}_{m-1,A} \cap \overline{\mathcal{F}}_{i-(m-1),A} \cap \mathcal{M}_{M-1-i,A_f} \cap \mathcal{F}_{1,dA}\right)$ | $P_{m,n}^{(m+n=k)}$ |
| $P_5$ | $P\left(\overline{\mathcal{F}}_{i,A} \cap \mathcal{M}_{M-i,A_f}\right)$ | $P_{m=0,n=k}^{(m+n=k)}$ |
| $P_6$ | $P\left(\overline{\mathcal{T}}_{j,A} \cap \mathcal{N}_{N-j,A_f}\right)$ | $P_{m=k,n=0}^{(m+n=k)}$ |
| $P_7$ | $P\left(\mathcal{T}_{n-1,A} \cap \overline{\mathcal{T}}_{j-(n-1),A} \cap \overline{\mathcal{T}}_{N-1-j,A_f} \cap \mathcal{T}_{1,dA}\right)$ | $P_{m,n}^{(m+n<k)}$ |
| $P_8$ | $P\left(\mathcal{F}_{m,A} \cap \overline{\mathcal{F}}_{i-m,A} \cap \overline{\mathcal{F}}_{M-i,A_f}\right)$ | $P_{m,n}^{(m+n<k)}$ |
| $P_9$ | $P\left(\mathcal{T}_{n,A} \cap \overline{\mathcal{T}}_{j-n,A} \cap \overline{\mathcal{T}}_{N-j,A_f}\right)$ | $P_{m,n}^{(m+n<k)}$ |
| $P_{10}$ | $P\left(\mathcal{F}_{m-1,A} \cap \overline{\mathcal{F}}_{i-(m-1),A} \cap \overline{\mathcal{F}}_{M-1-i,A_f} \cap \mathcal{F}_{1,dA}\right)$ | $P_{m,n}^{(m+n>k)}$ |
| $P_{11}$ | $P\left(\mathcal{F}_{0,A_s}\right)$ | $P_{m=0,n<k}^{(m+n<k)}$ |
| $P_{12}$ | $P\left(\mathcal{T}_{0,A_s}\right)$ | $P_{m<k,n=0}^{(m+n<k)}$ |

and an equivalent way to calculate $P\left(n \geq \hat{n}\right)$ is using

$$P\left(n \geq \hat{n}\right) = 1 - P\left(n < \hat{n}\right)$$
$$= 1 - \sum_{n=0}^{\min\left(N,\hat{n}-1\right)} \left[\sum_{m=0}^{\min\left(M,k-n-1\right)} P_{m,n}^{(m+n<k)}\right] - \sum_{n=\max\left(0,k-M\right)}^{\min\left(N,\hat{n}-1\right)} P_{k-n,n}^{(m+n=k)}. \quad (7.11)$$

When using either Eqs. (7.10) or (7.11), the case when both $m = 0$ and $n = 0$ must also be considered separately. This can occur only when warheads are left over (assuming $k \geq 1$). The resulting probability is

$$P_{m=0,n=0}^{(m+n<k)}\left(A_s\right) = P\left(\mathcal{F}_{0,A_s}\right) P\left(\mathcal{T}_{0,A_s}\right). \quad (7.12)$$

Viewing Eqs. (7.2)–(7.12), 12 "elemental" probabilities emerge that need to be determined regardless of the probability distributions used. Table 7.3 lists the 12 elemental probabilities and the corresponding conditions.

Allowing $P_{TR}$ and $(1 - P_{FTR})$ to vary with time means integrals will appear in the expressions for the elemental probabilities. To ease the appearance of the equations and to help formulate optimal control problems, states are defined as the integrals. The state $q$ represents the probability of encountering a single object. The state $x$ represents either the unconditional probability of attacking a single false target (for Scenarios 4 and 6) or the Poisson parameter for false target attacks (for Scenarios 1, 2, 3, 5, and 7). The state $y$

**Table 7.4:** Scenario 1 elemental probabilities and state definitions.

| | |
|---|---|
| $P_1$ | $= \frac{1}{A_B} P_{TR}(t)\, Q(t)\, dt$ |
| $P_2$ | $= e^{-x(t)} \frac{[x(t)]^m}{m!}$ |
| $P_3$ | $= [1 - y(t)]$ |
| $P_4$ | $= e^{-x(t)} \frac{[x(t)]^{(m-1)}}{(m-1)!} [1 - P_{FTR}(t)]\, Q(t)\, \alpha(t)\, dt$ |
| $P_5$ | $= e^{-x(t)}$ |
| $P_6$ | $= [1 - y(t)]$ |
| $P_7$ | $= \frac{1}{A_B} P_{TR}(t)\, Q(t)\, dt$ |
| $P_8$ | $= e^{-x(T)} \frac{[x(t)]^m}{m!}$ |
| $P_9$ | $= [1 - y(T)]$ |
| $P_{10}$ | $= e^{-x(T)} \frac{[x(t)]^{(m-1)}}{(m-1)!} [1 - P_{FTR}(t)]\, Q(t)\, \alpha(t)\, dt$ |
| $P_{11}$ | $= e^{-x(T)}$ |
| $P_{12}$ | $= [1 - y(T)]$ |
| $x(t)$ | $\equiv \int_0^t [1 - P_{FTR}(\tau)]\, Q(\tau)\, \alpha(\tau)\, d\tau$ |
| $y(t)$ | $\equiv \int_0^t \frac{1}{A_B} P_{TR}(\tau)\, Q(\tau)\, d\tau$ |

represents either the unconditional probability of attacking a single target (for Scenarios 1, 3, 4, 5, 6, and 7) or the Poisson parameter for target attacks (for Scenario 2).

Tables 7.4–7.10 contain two-part tables for each scenario containing the 12 elemental probabilities and corresponding state definitions. See [5] for detailed derivations. *Temporal* variables are used for Scenarios 1–4, Tables 7.4–7.7, assuming the mission ends at time $T$. *Spatial* variables are used for Scenarios 5–7, Tables 7.8–7.10, assuming the mission ends at radius $R$.

## 7.4 Optimal Control Formulation

A designer or mission planner wants the expected number of target attacks to be high and the expected number of false target attacks to be low. If $n$ is the actual number of target attacks during a mission and $m$ is the actual number of false target attacks, then one may want to maximize the probability of at least $\hat{n}$ target attacks, $P\left(n \geq \hat{n}\right)$, while constraining the probability of at least $\hat{m}$ false target attacks, $P\left(m \geq \hat{m}\right)$. Unfortunately due to the ROC, adjusting the sensor threshold to increase the number of target attacks also increases the number of false target attacks. Thus, the operator's objectives are competing, and a trade-off situation arises. The general optimization problem statement is

$$\begin{array}{rl} \text{Max} & P\left(n \geq \hat{n}\right) \\ \text{subject to} & P\left(m \geq \hat{m}\right) \leq b, \end{array}$$

where the upper bound $b$, threshold for target attacks $\hat{n}$, threshold for false target attacks $\hat{m}$, and mission duration $T$ are set by the designer or mission planner. The decision variable is $P_{TR}$.

**Table 7.5:** Scenario 2 elemental probabilities and state definitions.

$$
\begin{aligned}
P_1 &= e^{-y(t)}\frac{[y(t)]^{(n-1)}}{(n-1)!}P_{TR}(t)\,Q(t)\,\beta(t)\,dt\\
P_2 &= e^{-x(t)}\frac{[x(t)]^m}{m!}\\
P_3 &= e^{-y(t)}\frac{[y(t)]^n}{n!}\\
P_4 &= e^{-x(t)}\frac{[x(t)]^{(m-1)}}{(m-1)!}\left[1-P_{FTR}(t)\right]Q(t)\,\alpha(t)\,dt\\
P_5 &= e^{-x(t)}\\
P_6 &= e^{-y(t)}\\
P_7 &= e^{-y(T)}\frac{[y(t)]^{(n-1)}}{(n-1)!}P_{TR}(t)\,Q(t)\,\beta(t)\,dt\\
P_8 &= e^{-x(T)}\frac{[x(t)]^m}{m!}\\
P_9 &= e^{-y(T)}\frac{[y(t)]^n}{n!}\\
P_{10} &= e^{-x(T)}\frac{[x(t)]^{(m-1)}}{(m-1)!}\left[1-P_{FTR}(t)\right]Q(t)\,\alpha(t)\,dt\\
P_{11} &= e^{-x(T)}\\
P_{12} &= e^{-y(T)}
\end{aligned}
$$

$$
\begin{aligned}
x(t) &\equiv \int_0^t \left[1-P_{FTR}(\tau)\right]Q(\tau)\,\alpha(\tau)\,d\tau\\
y(t) &\equiv \int_0^t P_{TR}(\tau)\,Q(\tau)\,\beta(\tau)\,d\tau
\end{aligned}
$$

**Table 7.6:** Scenario 3 elemental probabilities and state definitions.

$$
\begin{aligned}
P_1 &= \sum_{j=n-1}^{N-1}\left\{\binom{N-1}{j}\binom{j}{n-1}[y(t)]^{n-1}[q(t)-y(t)]^{j-(n-1)}[1-q(t)]^{N-1-j}\right\}\\
&\quad\times\frac{N}{A_B}P_{TR}(t)\,Q(t)\,dt\\
P_2 &= e^{-x(t)}\frac{[x(t)]^m}{m!}\\
P_3 &= \sum_{j=n}^{N}\left\{\binom{N}{j}\binom{j}{n}[y(t)]^{n}[q(t)-y(t)]^{j-n}[1-q(t)]^{N-j}\right\}\\
P_4 &= e^{-x(t)}\frac{[x(t)]^{(m-1)}}{(m-1)!}\left[1-P_{FTR}(t)\right]Q(t)\,\alpha(t)\,dt\\
P_5 &= e^{-x(t)}\\
P_6 &= [1-y(t)]^N\\
P_7 &= \sum_{j=n-1}^{N-1}\left\{\binom{N-1}{j}\binom{j}{n-1}[y(t)]^{n-1}[q(t)-y(t)]^{j-(n-1)}\times\right.\\
&\quad\left.[1-q(t)-y(T)+y(t)]^{N-1-j}\right\}\frac{N}{A_B}P_{TR}(t)\,Q(t)\,dt\\
P_8 &= e^{-x(T)}\frac{[x(t)]^m}{m!}\\
P_9 &= \sum_{j=n}^{N}\left\{\binom{N}{j}\binom{j}{n}[y(t)]^{n}[q(t)-y(t)]^{j-n}[1-q(t)-y(T)+y(t)]^{N-j}\right\}\\
P_{10} &= e^{-x(T)}\frac{[x(t)]^{(m-1)}}{(m-1)!}\left[1-P_{FTR}(t)\right]Q(t)\,\alpha(t)\,dt\\
P_{11} &= e^{-x(T)}\\
P_{12} &= [1-y(T)]^N
\end{aligned}
$$

$$
\begin{aligned}
q(t) &\equiv \int_0^t \frac{1}{A_B}Q(\tau)\,d\tau\\
x(t) &\equiv \int_0^t \left[1-P_{FTR}(\tau)\right]Q(\tau)\,\alpha(\tau)\,d\tau\\
y(t) &\equiv \int_0^t \frac{1}{A_B}P_{TR}(\tau)\,Q(\tau)\,d\tau
\end{aligned}
$$

**Table 7.7:** Scenario 4 elemental probabilities and state definitions.

$$
\begin{aligned}
P_1 &= \textstyle\sum_{j=n-1}^{N-1} \left\{ \binom{N-1}{j}\binom{j}{n-1} [y(t)]^{n-1} [q(t) - y(t)]^{j-(n-1)} [1 - q(t)]^{N-1-j} \right\} \\
&\quad \times \tfrac{N}{A_B} P_{TR}(t)\, Q(t)\, dt \\
P_2 &= \textstyle\sum_{i=m}^{M} \left\{ \binom{M}{i}\binom{i}{m} [x(t)]^{m} [q(t) - x(t)]^{i-m} [1 - q(t)]^{M-i} \right\} \\
P_3 &= \textstyle\sum_{j=n}^{N} \left\{ \binom{N}{j}\binom{j}{n} [y(t)]^{n} [q(t) - y(t)]^{j-n} [1 - q(t)]^{N-j} \right\} \\
P_4 &= \textstyle\sum_{i=m-1}^{M-1} \left\{ \binom{M-1}{i}\binom{i}{m-1} [x(t)]^{m-1} [q(t) - x(t)]^{i-(m-1)} [1 - q(t)]^{M-1-i} \right\} \\
&\quad \times \tfrac{M}{A_B} [1 - P_{FTR}(t)]\, Q(t)\, dt \\
P_5 &= [1 - x(t)]^M \\
P_6 &= [1 - y(t)]^N \\
P_7 &= \textstyle\sum_{j=n-1}^{N-1} \left\{ \binom{N-1}{j}\binom{j}{n-1} [y(t)]^{n-1} [q(t) - y(t)]^{j-(n-1)} \right. \\
&\quad \left. \times [1 - q(t) - y(T) + y(t)]^{N-1-j} \right\} \tfrac{N}{A_B} P_{TR}(t)\, Q(t)\, dt \\
P_8 &= \textstyle\sum_{i=m}^{M} \left\{ \binom{M}{i}\binom{i}{m} [x(t)]^{m} [q(t) - x(t)]^{i-m} [1 - q(t) - x(T) + x(t)]^{M-i} \right\} \\
P_9 &= \textstyle\sum_{j=n}^{N} \left\{ \binom{N}{j}\binom{j}{n} [y(t)]^{n} [q(t) - y(t)]^{j-n} [1 - q(t) - y(T) + y(t)]^{N-j} \right\} \\
P_{10} &= \textstyle\sum_{i=m-1}^{M-1} \left\{ \binom{M-1}{i}\binom{i}{m-1} [x(t)]^{m-1} [q(t) - x(t)]^{i-(m-1)} \right. \\
&\quad \left. \times [1 - q(t) - x(T) + x(t)]^{M-1-i} \right\} \tfrac{M}{A_B} [1 - P_{FTR}(t)]\, Q(t)\, dt \\
P_{11} &= [1 - x(T)]^M \\
P_{12} &= [1 - y(T)]^N
\end{aligned}
$$

$$
\begin{aligned}
q(t) &\equiv \textstyle\int_0^t \tfrac{1}{A_B} Q(\tau)\, d\tau \\
x(t) &\equiv \textstyle\int_0^t \tfrac{1}{A_B} [1 - P_{FTR}(\tau)]\, Q(\tau)\, d\tau \\
y(t) &\equiv \textstyle\int_0^t \tfrac{1}{A_B} P_{TR}(\tau)\, Q(\tau)\, d\tau
\end{aligned}
$$

The problem can be formulated as an optimal control problem whose solution method is given in various textbooks. The notation used by Bryson and Ho [3] is used and the problem is posed in the standard form, which is

$$
\begin{aligned}
\text{Min} \quad & J[\mathbf{s}(t), \mathbf{u}(t)] = \phi[\mathbf{s}(T)] + \int_0^T L[\mathbf{s}(t), \mathbf{u}(t), t]\, dt \\
\text{Subject to} \quad & \dot{\mathbf{s}} = \mathbf{f}[\mathbf{s}(t), \mathbf{u}(t), t],\ \mathbf{s}(0) = 0.
\end{aligned}
$$

Since the standard form involves minimizing a cost functional $J$, the objective function [maximizing $P(n \geq \hat{n})$] is multiplied by minus one. The integrand $L$ then becomes the negative of the probability density function for $P(n \geq \hat{n})$. The states are put in a vector $\mathbf{s}(t)$. Initial conditions on the states are known. The vector $\mathbf{f}[\mathbf{s}(t), \mathbf{u}(t), t]$ specifies the dynamics. The control vector $\mathbf{u}(t)$, in this case, consists of the scalar decision variable $P_{TR}$. The performance objective $\phi$ is a function of the the final states.

The Pontryagin maximum principle [3] is invoked, which involves forming the Hamiltonian, $H$, given by

$$
H[\mathbf{s}(t), \mathbf{u}(t), t] = L[\mathbf{s}(t), \mathbf{u}(t), t] + \lambda^{\mathrm{T}}(t)\, \mathbf{f}[\mathbf{s}(t), \mathbf{u}(t), t], \tag{7.13}
$$

**Table 7.8:** Scenario 5 elemental probabilities and state definitions.

| | |
|---|---|
| $P_1$ | $= \sum_{j=n-1}^{N-1} \left\{ \binom{N-1}{j} \binom{j}{n-1} [y(r)]^{n-1} \left[q_y(r) - y(r)\right]^{j-(n-1)} \left[1 - q_y(r)\right]^{N-1-j} \right\}$ $\times N P_{TR}(r) \frac{r}{\sigma_y^2} e^{-\frac{r^2}{2\sigma_y^2}} dr$ |
| $P_2$ | $= e^{-x(r)} \frac{[x(r)]^m}{m!}$ |
| $P_3$ | $= \sum_{j=n}^{N} \left\{ \binom{N}{j} \binom{j}{n} [y(r)]^{n} \left[q_y(r) - y(r)\right]^{j-n} \left[1 - q_y(r)\right]^{N-j} \right\}$ |
| $P_4$ | $= e^{-x(r)} \frac{[x(r)]^{(m-1)}}{(m-1)!} [1 - P_{FTR}(r)] 2\pi r \alpha(r) dr$ |
| $P_5$ | $= e^{-x(r)}$ |
| $P_6$ | $= [1 - y(r)]^N$ |
| $P_7$ | $= \sum_{j=n-1}^{N-1} \left\{ \binom{N-1}{j} \binom{j}{n-1} [y(r)]^{n-1} \left[q_y(r) - y(r)\right]^{j-(n-1)} \times \right.$ $\left. \left[1 - q_y(r) - y(R) + y(r)\right]^{N-1-j} \right\} N P_{TR}(r) \frac{r}{\sigma_y^2} e^{-\frac{r^2}{2\sigma_y^2}} dr$ |
| $P_8$ | $= e^{-x(R)} \frac{[x(r)]^m}{m!}$ |
| $P_9$ | $= \sum_{j=n}^{N} \left\{ \binom{N}{j} \binom{j}{n} [y(r)]^{n} \left[q_y(r) - y(r)\right]^{j-n} \left[1 - q_y(r) - y(R) + y(r)\right]^{N-j} \right\}$ |
| $P_{10}$ | $= e^{-x(R)} \frac{[x(r)]^{(m-1)}}{(m-1)!} [1 - P_{FTR}(r)] 2\pi r \alpha(r) dr$ |
| $P_{11}$ | $= e^{-x(R)}$ |
| $P_{12}$ | $= [1 - y(R)]^N$ |
| $q_y(r) \equiv \int_0^r \frac{\rho}{\sigma_y^2} e^{-\frac{\rho^2}{2\sigma_y^2}} d\rho$ | |
| $x(r) \equiv \int_0^r [1 - P_{FTR}(\rho)] 2\pi \rho \alpha(\rho) d\rho$ | |
| $y(r) \equiv \int_0^r P_{TR}(\rho) \frac{\rho}{\sigma_y^2} e^{-\frac{\rho^2}{2\sigma_y^2}} d\rho$ | |

where $\lambda$ is the vector of costates. Note that $T$ is mission duration and T is the transpose operator. The costate differential equations are

$$\dot{\lambda}^{\mathrm{T}} = -\frac{\partial H}{\partial \mathbf{s}} = -\frac{\partial L}{\partial \mathbf{s}} - \lambda^{\mathrm{T}} \frac{\partial \mathbf{f}}{\partial \mathbf{s}}. \tag{7.14}$$

Finally, the optimal control is derived using the first-order necessary condition

$$\frac{\partial H}{\partial \mathbf{u}} = \mathbf{0}. \tag{7.15}$$

The differential equations from this formulation are coupled, so closed-form solutions do not exist. Instead, the state and costate equations must be numerically integrated. Two-point boundary value problems of this form can be solved using a "shooting method" or a discrete time, gradient-based solver.

### 7.4.1 Sample Formulation

To clarify the approach, a sample formulation is provided. The following are where the various parameters originate:

**Table 7.9:** Scenario 6 elemental probabilities and state definitions.

| | |
|---|---|
| $P_1$ | $= \sum_{j=n-1}^{N-1} \binom{N-1}{j}\binom{j}{n-1} [y(r)]^{n-1} \left[q_y(r) - y(r)\right]^{j-(n-1)} \left[1 - q_y(r)\right]^{N-1-j}$ $\times N P_{TR}(r) \frac{r}{\sigma_y^2} e^{-\frac{r^2}{2\sigma_y^2}} dr$ |
| $P_2$ | $= \sum_{i=m}^{M} \binom{M}{i}\binom{i}{m} [x(r)]^m [q_x(r) - x(r)]^{i-m} [1 - q_x(r)]^{M-i}$ |
| $P_3$ | $= \sum_{j=n}^{N} \binom{N}{j}\binom{j}{n} [y(r)]^n \left[q_y(r) - y(r)\right]^{j-n} \left[1 - q_y(r)\right]^{N-j}$ |
| $P_4$ | $= \sum_{i=m-1}^{M-1} \binom{M-1}{i}\binom{i}{m-1} [x(r)]^{m-1} [q_x(r) - x(r)]^{i-(m-1)} [1 - q_x(r)]^{M-1-i}$ $\times M [1 - P_{FTR}(r)] \frac{r}{\sigma_x^2} e^{-\frac{r^2}{2\sigma_x^2}} dr$ |
| $P_5$ | $= [1 - x(r)]^M$ |
| $P_6$ | $= [1 - y(r)]^N$ |
| $P_7$ | $= \sum_{j=n-1}^{N-1} \binom{N-1}{j}\binom{j}{n-1} [y(r)]^{n-1} \left[q_y(r) - y(r)\right]^{j-(n-1)} \times$ $\left[1 - q_y(r) - y(R) + y(r)\right]^{N-1-j} N P_{TR}(r) \frac{r}{\sigma_y^2} e^{-\frac{r^2}{2\sigma_y^2}} dr$ |
| $P_8$ | $= \sum_{i=m}^{M} \binom{M}{i}\binom{i}{m} [x(r)]^m [q_x(r) - x(r)]^{i-m} [1 - q_x(r) - x(R) + x(r)]^{M-i}$ |
| $P_9$ | $= \sum_{j=n}^{N} \binom{N}{j}\binom{j}{n} [y(r)]^n \left[q_y(r) - y(r)\right]^{j-n} \left[1 - q_y(r) - y(R) + y(r)\right]^{N-j}$ |
| $P_{10}$ | $= \sum_{i=m-1}^{M-1} \binom{M-1}{i}\binom{i}{m-1} [x(r)]^{m-1} [q_x(r) - x(r)]^{i-(m-1)} \times$ $[1 - q_x(r) - x(R) + x(r)]^{M-1-i} M [1 - P_{FTR}(r)] \frac{r}{\sigma_x^2} e^{-\frac{r^2}{2\sigma_x^2}} dr$ |
| $P_{11}$ | $= [1 - x(R)]^M$ |
| $P_{12}$ | $= [1 - y(R)]^N$ |
| $q_x(r) \equiv \int_0^r \frac{\rho}{\sigma_x^2} e^{-\frac{\rho^2}{2\sigma_x^2}} d\rho$ | $q_y(r) \equiv \int_0^r \frac{\rho}{\sigma_y^2} e^{-\frac{\rho^2}{2\sigma_y^2}} d\rho$ |
| $x(r) \equiv \int_0^r [1 - P_{FTR}(\rho)] \frac{\rho}{\sigma_x^2} e^{-\frac{\rho^2}{2\sigma_x^2}} d\rho$ | |
| $y(r) \equiv \int_0^r P_{TR}(\rho) \frac{\rho}{\sigma_y^2} e^{-\frac{\rho^2}{2\sigma_y^2}} d\rho$ | |

- Designer or mission planner: $\hat{m}, \hat{n}, b, k, T, R, Q$,
- Order of battle intelligence: $\alpha, \beta, N, M, \sigma_x, \sigma_y, A_B$,
- Performance specifications: $c$.

Let $\hat{m} = \hat{n} = 1$, and assume $T$, $b$, and $Q$ are designated. In addition, assume both the number of targets $N$ and the number of false targets $M$ are both greater than the number of warheads $k$. At this point, the actual values for $T$, $b$, $Q$, $N$, $M$, and $k$ are not needed and can remain parameters. The optimization problem becomes

$$\begin{array}{rl} \text{Max} & P(n \geq 1) \\ \text{subject to} & P(m \geq 1) \leq b. \end{array}$$

**Table 7.10:** Scenario 7 elemental probabilities and state definitions.

| | |
|---|---|
| $P_1$ | $= P_{TR}(r)\frac{r}{\sigma_y^2}e^{-\frac{r^2}{2\sigma_y^2}}dr$ |
| $P_2$ | $= e^{-x(r)}\frac{[x(r)]^m}{m!}$ |
| $P_3$ | $= [1 - y(r)]$ |
| $P_4$ | $= e^{-x(r)}\frac{[x(r)]^{(m-1)}}{(m-1)!}[1 - P_{FTR}(r)]2\pi r\alpha(r)dr$ |
| $P_5$ | $= e^{-x(r)}$ |
| $P_6$ | $= [1 - y(r)]$ |
| $P_7$ | $= P_{TR}(r)\frac{r}{\sigma_y^2}e^{-\frac{r^2}{2\sigma_y^2}}dr$ |
| $P_8$ | $= e^{-x(R)}\frac{[x(r)]^m}{m!}$ |
| $P_9$ | $= [1 - y(R)]$ |
| $P_{10}$ | $= e^{-x(R)}\frac{[x(r)]^{(m-1)}}{(m-1)!}[1 - P_{FTR}(r)]2\pi r\alpha(r)dr$ |
| $P_{11}$ | $= e^{-x(R)}$ |
| $P_{12}$ | $= [1 - y(R)]$ |
| $x(r)$ | $\equiv \int_0^r [1 - P_{FTR}(\rho)]2\pi\rho\alpha(\rho)d\rho$ |
| $y(r)$ | $\equiv \int_0^r P_{TR}(\rho)\frac{\rho}{\sigma_y^2}e^{-\frac{\rho^2}{2\sigma_y^2}}d\rho$ |

Using Eqs. (7.11) and (7.10) gives

$$P(n \geq 1) = 1 - P(n = 0)$$
$$= 1 - \sum_{m=0}^{k-1} P_{m,0}^{(m+n<k)} - P_{k,0}^{(m+n=k)} \tag{7.16}$$

and

$$P(m \geq 1) = 1 - P(m = 0)$$
$$= 1 - \sum_{n=0}^{k-1} P_{0,n}^{(m+n<k)} - P_{0,k}^{(m+n=k)}. \tag{7.17}$$

Substituting elemental probabilities gives

$$P(n \geq 1) = 1 - P_{11}P_{12} - \sum_{m=1}^{k-1}\int_0^T P_{10}P_{12} - \int_0^T P_4P_6 \tag{7.18}$$

and

$$P(m \geq 1) = 1 - P_{11}P_{12} - \sum_{n=1}^{k-1}\int_0^T P_7P_{11} - \int_0^T P_1P_5. \tag{7.19}$$

Bryson and Ho notation becomes

$$\phi[\mathbf{s}(T)] = P_{11}P_{12}, \tag{7.20}$$

$$L\left[\mathbf{s}\left(t\right),\mathbf{u}\left(t\right)\right]dt = P_4 P_6 + \sum_{m=1}^{k-1} P_{10} P_{12}, \tag{7.21}$$

$$\mathbf{u}\left(t\right) = P_{TR}\left(t\right), \tag{7.22}$$

$$\mathbf{s}\left(t\right) = \left[q\left(t\right)\ x\left(t\right)\ y\left(t\right)\ z\left(t\right)\right]^{\mathrm{T}}, \tag{7.23}$$

where $q(t)$, $x(t)$, and $y(t)$ are defined once a scenario is selected, and

$$z\left(t\right) = 1 - P_{11} P_{12} - \int_0^T \left( P_1 P_5 + \sum_{n=1}^{k-1} P_7 P_{11} \right). \tag{7.24}$$

For example, if Scenario 2 elemental probabilities are used and $Q$ is constant, then

$$x\left(t\right) = \int_0^t \left[ \frac{P_{TR}\left(\tau\right)}{\left(1-c\right) P_{TR}\left(\tau\right) + c} \right] Q\alpha\left(\tau\right) d\tau, \tag{7.25}$$

$$y\left(t\right) = \int_0^t P_{TR}\left(\tau\right) Q\beta\left(\tau\right) d\tau, \tag{7.26}$$

$$\begin{aligned} z\left(t\right) = 1 - e^{-x(t)} e^{-y(t)} - \int_0^t \Bigg\{ & e^{-y(\tau)} \frac{\left[y\left(\tau\right)\right]^{(n-1)}}{(n-1)!} P_{TR}\left(\tau\right) Q\beta\left(\tau\right) e^{-x(\tau)} \\ & + \sum_{n=1}^{k-1} e^{-y(t)} \frac{\left[y\left(\tau\right)\right]^{(n-1)}}{(n-1)!} P_{TR}\left(\tau\right) Q\beta\left(\tau\right) e^{-x(t)} \Bigg\} d\tau, \end{aligned} \tag{7.27}$$

$$\phi\left[\mathbf{s}\left(T\right)\right] = e^{-x(T)} e^{-y(T)}, \tag{7.28}$$

$$\begin{aligned} L\left[\mathbf{s}\left(t\right),\mathbf{u}\left(t\right)\right] = {} & e^{-x(t)} \frac{\left[x\left(t\right)\right]^{(m-1)}}{(m-1)!} \left[ \frac{P_{TR}\left(t\right)}{\left(1-c\right) P_{TR}\left(t\right) + c} \right] Q\alpha\left(t\right) e^{-y(t)} \\ & + \sum_{m=1}^{k-1} e^{-x(T)} \frac{\left[x\left(t\right)\right]^{(m-1)}}{(m-1)!} \left[ \frac{P_{TR}\left(t\right)}{\left(1-c\right) P_{TR}\left(t\right) + c} \right] Q\alpha\left(t\right) e^{-y(T)}. \end{aligned} \tag{7.29}$$

## 7.5 Applications

An analysis was performed in [5] for some computationally tractable scenarios when $k = 1$ or $k = \infty$. The objective of the analysis was to bound the problem and verify numerical solutions. The case where $k = 1$ represents a munition. Ironically, the case where $k = \infty$ can be used to mathematically evaluate a zero-warhead sensor craft. This follows from the fact that a sensor craft broadcasts target coordinates assuming a shooter exists to destroy the targets. A sensor craft's mission does not terminate because it has no more warheads. It continues to search and classify until it is sent home (or shot down). Thus mathematically, a sensor craft is equivalent to a vehicle having an infinite number of warheads. Four conclusions of the analysis follow:

1. For Scenarios 1 and 2 with $k = 1$ and no constraint on $P\left(m \geq \hat{m}\right)$, if $T$, $Q$, $\alpha$, and $\beta$ are constant, the unbounded $P_{TR}^*$ is monotonically increasing with time. Further, the constraint $P_{TR} \leq 1$ becomes active before the end and remains active until time $T$. This represents a "go for broke" tactic in the end game. This strategy of starting conservative then gradually increasing aggressiveness until the end, where an all-out effort is tried, is common in game theory. Without a constraint on false target attacks, the system has nothing to lose by declaring all objects as targets near the end.

2. For Scenarios 1, 2, and 7 with $k = \infty$, $\hat{n} = 1$, and no constraint on $P\left(m \geq \hat{m}\right)$, if $T$ (or $R$), $Q$, $\alpha$, and $\beta$ are constant, the bounded $P_{TR}^* = P_{TR_{max}}^* = 1$. This is not a practical solution, since all false targets would be declared targets and attacked. With an infinite amount of warheads and no constraint on false target attacks, there is no reason not to attack every object encountered.

3. For Scenarios 1 and 2 with $k = \infty$ and $P\left(m \geq \hat{m}\right) = b$, if $T$, $Q$, $\alpha$, and $\beta$ are constant, then $P_{TR}^*$ is constant for all time. When target and false target densities remain fixed throughout the search effort and the amount of warheads is infinite, the problem becomes an infinite horizon problem, where optimal solutions are constant.

4. For Scenario 7 with $k = \infty$ and $P\left(m \geq \hat{m}\right) = b$, if $R$ and $\alpha$ are constant, then $P_{TR}^*$ is monotonically decreasing with radius. As the radius increases, the amount of expected false target encounters in an annulus *increases*, and the probability of encountering the target in an annulus *decreases*. With infinite warheads, the focus becomes meeting the constraint on false target attacks. Since the amount of expected false target encounters in an annulus *increases* with radius, it makes intuitive sense to tighten the operating point on the ROC curve. That is, decrease $P_{TR}$ as radius increases.

Consider the probability of at least one target attack, $P\,(n \geq 1)$, and the probability of at least one false target attack, $P\,(m \geq 1)$, for effectiveness measures. Having $\hat{n} = \hat{m} = 1$ allows one to compare the munition, tactical UAV, and sensor craft problems. Thus, the sensitivity of effectiveness to the number of warheads can be evaluated. By varying $b$, the sensitivity of effectiveness to the upper bound on false target attacks can be evaluated. Finally, by examining a variety of battle space scenarios, the effect of the target-to-false-target ratio being fixed or a function of sensor footprint location can be evaluated.

### 7.5.1 Numerical Solution Method

The following is a synopsis of Bryson's gradient algorithm [2]. First, the continuous state equations are put into discrete form using Euler's method. The problem is then put into Mayer form, where the state vector is augmented by one state representing the cumulative sum of $L$ to step $i$. The problem statement then becomes

$$
\begin{aligned}
\text{Min} \quad & \phi\left[\mathbf{s}\left(\mathrm{N}\right)\right] \\
\text{subject to} \quad & \mathbf{s}\left(i+1\right) = \mathbf{f}\left[\mathbf{s}\left(i\right), \mathbf{u}\left(i\right), i\right], \qquad i = 0, \ldots, \mathrm{N}-1, \\
& \mathbf{s}\left(0\right) = \mathbf{s}_0, \\
& \psi\left[\mathbf{s}\left(\mathrm{N}\right)\right] = 0,
\end{aligned}
$$

where N is the number of time or distance steps, as opposed to $N$, which is the number of targets. $\psi$ represents terminal constraints and is some function of the terminal states. In this case, $\psi\left[\mathbf{s}\left(\mathrm{N}\right)\right] = z\left(\mathrm{N}\right) - b$. The Mayer form is equivalent to the Bolza form; however, the Mayer form yields simpler expressions and is easier to code for numerical solutions. MATLAB's *fmincon* routine is used, so one does not have to guess step size. The values for $\mathbf{f}$, $\mathbf{f_s}$, $\mathbf{f_u}$, $\phi_\mathbf{s}$, and $\psi_\mathbf{s}$ at any time or distance step must be provided in a subroutine. The algorithm then proceeds as follows:

- Guess $\mathbf{u}(i)$ for $i = 0, \ldots, \mathrm{N}-1$.
- Forward propagate the state equations from $\mathbf{s}(0)$ using $\mathbf{u}(i)$ and store $\mathbf{s}(i)$.
- Evaluate $\phi$ and $\psi$ and set $\lambda^\phi(\mathrm{N}) = \phi_\mathbf{s}^\mathrm{T}$, $\lambda^\psi(\mathrm{N}) = \psi_\mathbf{s}^\mathrm{T}$.
- Backward propagate and store the response sequences $H_\mathbf{u}^\phi(i)$ and $H_\mathbf{u}^\psi(i)$:

$$\begin{aligned} \lambda^\phi(i) &= \mathbf{f}_\mathbf{s}^\mathrm{T}(i)\,\lambda^\phi(i+1), \qquad i = \mathrm{N}-1, \ldots, 0, \\ \lambda^\psi(i) &= \mathbf{f}_\mathbf{s}^\mathrm{T}(i)\,\lambda^\psi(i+1), \\ \left[H_\mathbf{u}^\phi(i)\right]^\mathrm{T} &= \mathbf{f}_\mathbf{u}^\mathrm{T}(i)\,\lambda^\phi(i+1), \\ \left[H_\mathbf{u}^\psi(i)\right]^\mathrm{T} &= \mathbf{f}_\mathbf{u}^\mathrm{T}(i)\,\lambda^\psi(i+1). \end{aligned}$$

The process is repeated until the terminal constraint error and the gradient sequence are negligibly small. The problem becomes a parameter optimization involving $n_c \times \mathrm{N}$ decision variables, where $n_c$ is the number of control variables.

### 7.5.2 Scenario 1 Example

Scenario 1 is a fixed-parameter problem formulation that has several advantages. The operator simply sets $P_{TR}$ at the beginning of the mission. No sophisticated scheduling software or hardware is needed. Finding the optimal $P_{TR}$ can be accomplished by examining a plot or table. For example, in Scenario 1 with $k = 1$, the effectiveness measures become

$$P(n \geq 1) = P_{TR}\frac{\left[1 - e^{-(1-P_{FTR})\alpha QT}\right]}{(1 - P_{FTR})\alpha A_B} \tag{7.30}$$

and

$$P(m \geq 1) = 1 - P_{TR}\frac{\left[1 - e^{-(1-P_{FTR})\alpha QT}\right]}{(1 - P_{FTR})\alpha A_B} - \left(1 - \frac{P_{TR}QT}{A_B}\right)e^{-(1-P_{FTR})\alpha QT}. \tag{7.31}$$

Eq. (7.1) can be used for the ROC model with $c = 100$. Assume the munition covers $A_B$ by time $T$. Thus, $QT = A_B$. One of three mutually exclusive events is possible. The munition either attacks a target, attacks a false target, or attacks nothing and self-destructs. Fig. 7.4 shows the outcome probabilities versus $P_{TR}$ when $c = 100$, $\alpha Q = 50$ (1/hr), and $T = 0.5$ hr.

If the constraint on a false target attack is removed, Fig. 7.4 shows that the best unconstrained solution is $P_{TR_u}^* = 0.723$ with a corresponding $P(n \geq 1)^* = 0.535$ and $P(m \geq 1)^* = 0.318$. If $P(m \geq 1)$ is bounded by $b = 0.2$, the best constrained solution is $P_{TR}^* = 0.563$ with a corresponding $P(n \geq 1)^* = 0.483$. In Fig. 7.4, the outcome probability functions are smooth and well-behaved. The function for $P(n \geq 1)$ has one

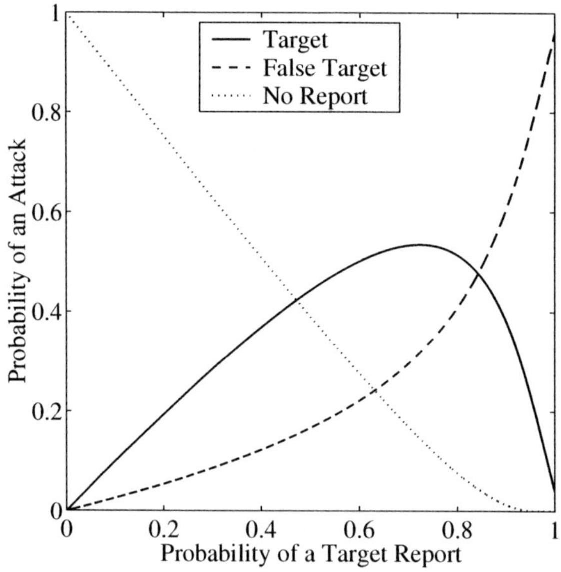


**Figure 7.4:** Outcome probabilities versus probability of a target report. Scenario 1 with constant parameters, $c = 100$, $\alpha Q = 50$ (1/hr), and $T = 0.5$ (hr).

peak. The function for $P(m \geq 1)$ is monotonically increasing, so any constrained solution will be unique. Finally, the function for the probability of no attack always starts at (0, 1), then monotonically decreases. Since the three outcome probabilities are mutually exclusive in a munition problem, they sum to one for each $P_{TR}$ considered.

The problem can be solved for a number of $b$ values and a plot of $P_{TR}^*$ versus $b$ can be generated. If $b \geq 0.318$, the unconstrained solution $P_{TR_u}^* = 0.723$ is used. Fig. 7.5 illustrates the sensitivity of the solution to changes in $b$. Fig. 7.5 is the plot an operator needs to set $P_{TR}$ at the beginning of a mission, assuming $b$ has been determined by the commander.

Fixed-parameter problems are readily solvable and simple to employ operationally. Improvement when parameters are allowed to vary is now examined. The case where $P_{TR}$ is allowed to vary while $Q$ remains fixed is considered. Having a fixed area coverage rate means a single ROC curve is used. Moving along a single ROC curve is accomplished by adjusting the sensor threshold, which is tantamount to adjusting $P_{TR}$.

Consider Scenario 1 with $k = 1$ and Eq. (7.1) for the ROC model with $c = 100$. Assume the munition covers $A_B$ by time $T$. Thus, $QT = A_B$. The optimal control problem can be solved for a number of $b$ values. Fig. 7.6 shows optimal $P_{TR}$ schedules for three constraint values. Analysis says the optimal $P_{TR}$ will increase over time when the constraint on false target attacks is removed. Fig. 7.6 shows the optimal $P_{TR}$ increases over time even when a constraint on false target attacks exists. The increases diminish as constraint values are tightened (lowered).

The problem can be solved for a number of constraint values, and the objective function values can be compared to those obtained using fixed parameters. In Fig. 7.7, there

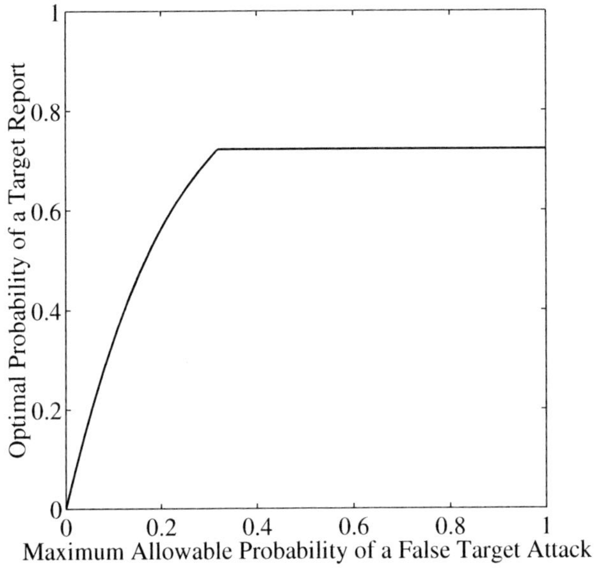


**Figure 7.5:** Optimal probability of a target report versus maximum allowable probability of a false target attack, $b$. Scenario 1 with constant parameters, $c = 100$, $\alpha Q = 50$ (1/hr), and $T = 0.5$ (hr).

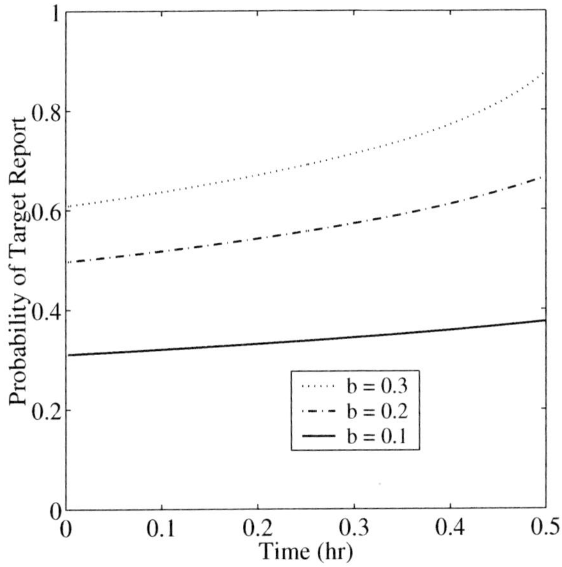


**Figure 7.6:** Optimal probability of a target report versus time. Scenario 1 with $c = 100$, $\alpha Q = 50$ (1/hr), and $T = 0.5$ (hr).

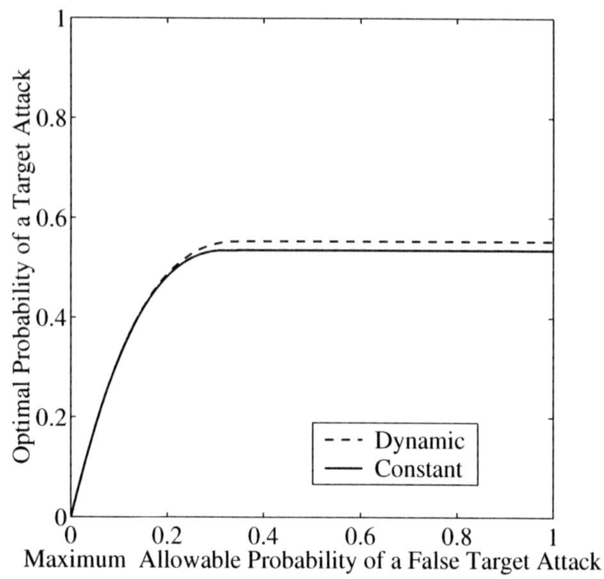


**Figure 7.7:** Optimal probability of a target attack versus maximum allowable probability of a false target attack. Scenario 1 with $c = 100$, $\alpha Q = 50$ (1/hr), and $T = 0.5$ (hr).

is negligible improvement for constraint values below 0.15 and marginal improvement (up to 3.4%) for constraint values above 0.15. Depending on the application, this improvement may or may not be worth the added complexity.

Analysis shows that the optimal $P_{TR}$ is constant when $k = \infty$ and, for Scenario 1, is given by

$$P_{TR}^{*}(t) = \frac{-c \ln[1 - b]}{(1 - c)\ln[1 - b] + \alpha Q T}. \tag{7.32}$$

Now consider the effect of multiple warheads. Fig. 7.8 shows the trade space for $k = 1, 2, 6, \infty$ and $\hat{m} = \hat{n} = 1$. The results for $k = \infty$ come analytically from Eq. (7.32). The $k = 1$ results do not exactly conform to other $k$ values, which are bound by $k = \infty$. The munition problem has three mutually exclusive outcomes, whereas the outcomes for the $k > 1$ problems are not mutually exclusive. Nonetheless, Fig. 7.8 can be used by commanders to make decisions about constraint values and number of warheads. For $b < 0.20$, there is not much to be gained by adding warheads. In this single-target scenario, there are more false targets present, which means the probability of the first attack being a false target attack is relatively high. Thus it does not matter how many warheads there are. The constraint is met by the first attack. For $b < 0.20$, one can simply use the constant closed-form $P_{TR}$ given by Eq. (7.32). In Fig. 7.8, it is clear the benefits of more than six warheads is negligible throughout most of the range of $b$. These types of insights are possible using the techniques and equations provided.

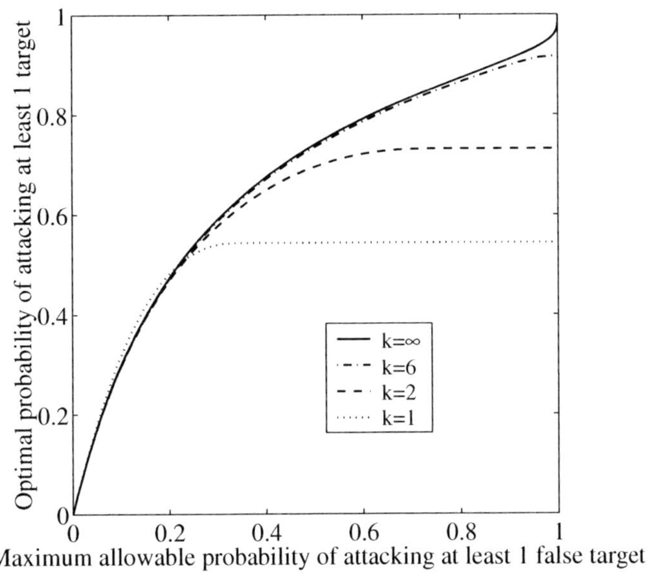


**Figure 7.8:** Optimal $P(n \geq 1)$ versus $b$. Scenario 1 with $c = 100$, $\alpha Q = 50$ (1/hr), and $T = 0.5$ (hr).

## 7.6 Summary

The objective of this chapter was to analytically quantify the operational effectiveness of airborne systems performing search-and-destroy missions in uncertain environments. Two types of search patterns were considered. The first type was the linear search pattern, which can be used to cover any linear-symmetric battle space. The second type involved concentric annuli emanating from the origin, which can be used to cover a circular battle space. Uniform, Poisson, and normal probability distributions were used to model target and false target encounters. Expressions needed to calculate the probability of at least $\hat{n}$ target attacks, $P\left(n \geq \hat{n}\right)$, and the probability of at least $\hat{m}$ false target attacks, $P\left(m \geq \hat{m}\right)$, were given for seven battle space scenarios.

Sensor performance, in terms of a system's ability to autonomously distinguish between targets and false targets, came from a ROC curve. A representative ROC curve model was used, whereby a parameter $c$ dictated the locus of operating points.

Optimal control problems were formulated to maximize $P\left(n \geq \hat{n}\right)$ subject to $P\left(m \geq \hat{m}\right) \leq b$, where $b$ was set by the user. The decision variable was $P_{TR}$. Due to coupling of the state and costate equations, the two point boundary value problems were solved numerically using a gradient algorithm. Analytic conclusions were provided for the munition and sensor craft problems for Scenarios 1, 2, and 7.

The results apply to search and attack vehicles with 0, 1, or $k$ warheads. Although the real motivation for this analysis comes from autonomous air operations, the results can be applied to manned, land-based, marine-based, and space-based systems. They can be applied in the design phase or operational phase of a system. In the design phase, the results can be used to develop system requirements, compare competing designs, or verify simulation results. Parameter sensitivity studies can be done to determine the cost-effectiveness of improving sensors, warheads, or vehicle performance. The mathematical framework enables sound systems engineering. In the operational phase, tactical decisions in the form of setting operating points and determining the number of warheads are possible using the results presented. Summary plots of $P\left(n \geq \hat{n}\right)$ versus $P\left(m \geq \hat{m}\right)$ that include warhead effects give mission planners a complete picture of the trade space. Planners can look at one figure and see the effects of changing constraint level or changing number of warheads.

## Bibliography

[1] *Air Force Basic Doctrine Document* 1, *AFM* 1-1. Technical report. United States Air Force, Sept. 1997.

[2] A. Bryson. *Dynamic Optimization.* Addison-Wesley, Reading, MA, 1999.

[3] A. Bryson and Y. Ho. *Applied Optimal Control.* Ginn and Company, New York, 1969.

[4] D. D. Decker. *Decision Factors for Cooperative Multiple Warhead UAV Target Classification and Attack.* Ph.D. thesis, Air Force Institute of Technology, Sept. 2004.

[5] B. A. Kish. *Establishment of a System Operating Characteristic for Autonomous Wide Area Search Vehicles.* Ph.D. thesis, Air Force Institute of Technology, Sept. 2005.

Appendix A

# MultiUAV Simulation

*Steven Rasmussen*

*I do not fear computers. I fear the lack of them.*
—Isaac Asimov

## A.1 MultiUAV2 Background

MultiUAV2 is a MATLAB\SIMULINK\C++ based multiple Unmanned Aerial Vehicle (UAV) cooperative control simulation. Construction of MultiUAV2[3] satisfies the need for a simulation environment that researchers can use to develop, implement, and analyze cooperative control algorithms. Since the purpose of MultiUAV2 is to make cooperative control research accessible to researchers, it was constructed primarily using MATLAB and SIMULINK. Simulation functions that require faster execution and do not need to be modified by researchers are programmed in C++. For example, the cooperative control algorithms are coded in MATLAB scripts and the vehicle dynamics are coded in C++.

The simulation is implemented in a hierarchical manner with intervehicle communications explicitly modeled. During construction of MultiUAV2, issues concerning memory usage and functional encapsulation were addressed. MultiUAV2 includes plotting tools and links to an external program for postsimulation analysis [4]. Each of the vehicle simulations include six-degree-of-freedom (6DOF) dynamics and embedded flight software (EFS). The EFS consists of a collection of managers that control situational awareness and responses of the vehicles. Managers included in the simulation are Sensor, Target, Cooperation, Route, and Weapons; see Fig. A.1.

During the simulation, vehicles fly predefined search trajectories until a target is encountered. Each vehicle has a mission sensor footprint that defines the sensor's field of

[3]The original MultiUAV was released to the public in 2002. In 2004 a new version of the simulation, MultiUAV2, was released.

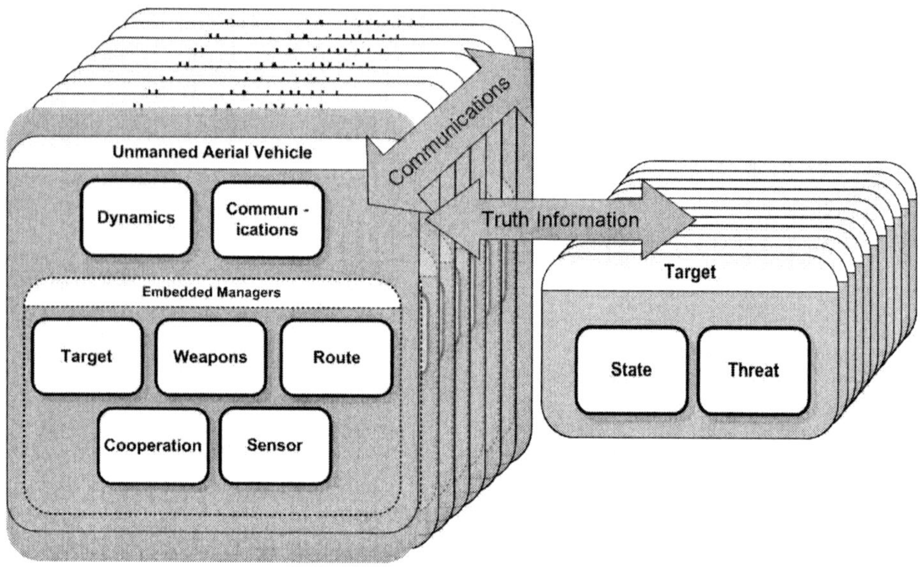


**Figure A.1:** MultiUAV2 top level organization.

view. Target positions either are set randomly or can be specified by the user. When a target position is inside a vehicle's sensor footprint, that vehicle runs a sensor simulation and sends the results to the other vehicles. With actions assigned by the selected cooperative control algorithm, the vehicles prosecute the known targets. The default settings of the simulation cause the vehicles to be destroyed when they attack a target. During prosecution of the targets the vehicles generate their own minimum-time paths to accomplish tasks. The simulation takes place in a three-dimensional environment, but all the path planning is for a constant altitude, i.e., two dimensions. Once each vehicle has finished its assigned tasks it returns to its predefined search pattern path. The simulation continues until it is stopped or the preset simulation run time has elapsed.

## A.2 Simulated Mission

In MultiUAV2 the mission is made up of the *participants*, e.g., UAVs and targets, the *mission objectives*, e.g., find and destroy the targets, and the *tactics*, i.e., how the UAVs meet the mission objectives. The nominal mission[4] in MultiUAV2 is a wide area search munition (WASM) mission. MultiUAV2's WASM mission objective is to find and destroy all of the targets. The UAVs search a predefined area for targets. If a suspected target is found, the UAVs must classify it to make sure that it is a target. Once a target is classified the vehicles attack it and then verify that it has been killed.

The WASM mission is implemented in MultiUAV2 by giving the UAVs the capability to perform tasks such as search, classify, attack, and verify destruction, based on the perceived

[4]The nominal mission is the default mission supplied with the released version of MultiUAV2.

state of the target. To do this, the UAVs must keep track of the state of the targets and cooperate to determine which UAV performs which task. Target state functions are used to track the target states shown in Fig. 3.2. The functions are implemented on each vehicle, for each target, to make it possible for each vehicle to keep track of the state of all of the targets. Based on the target's current state, the vehicle determines which task is necessary to transition the target to the next state. Algorithms to compute paths are defined for each task type. UAVs generate paths for all the current tasks and communicate the information required for cooperation to the other UAVs. The UAVs run their cooperation algorithms and implement their assignments.

To modify existing or add new missions, one can change the participants, mission objective, or tactics. To change the participants, new targets (see section A.3.1), and vehicles (see section A.3.2) can be added to the simulation or existing ones can be modified. The mission objectives and tactics are encoded in the target states and the task algorithms. The target states can be modified to cause the UAVs to perform different tasks or tasks in a different order and thus perform different missions. New task algorithms can be added or existing algorithms can be modified to cause the vehicles to perform tasks in a particular manner.

## A.3 Organization of MultiUAV2

MultiUAV2 is organized into two major top-level blocks, Vehicles and Targets; see Fig. A.2. The other blocks at the top level, Initialization, DataForPlotting, and Simulation Control, call functions to initialize the simulation, save simulation data for plotting, and control the simulation's execution. Nominally, the top-level blocks contain the subblocks and connections required to implement the simulation of vehicles and targets; see Figs. A.3–A.7. Information flow between the vehicles is facilitated with a *communication simulation*; see section A.3.4. Information flow between blocks within each vehicle is implemented using SIMULINK *wires* and, sparingly, global MATLAB memory. To facilitate simulation truth data flow between the objects in the simulation a truth message passing mechanism is used; see section A.3.4.

By default, 8 vehicles and 10 targets are implemented in MultiUAV2. By changing global parameters, the number of targets and vehicles used in a simulation can be decreased. The number of vehicles and targets can be increased by adding more blocks and making changes to global parameters. By default, all the vehicles in this simulation are homogeneous and share the same simulation structure. The same is true for the targets. Therefore, to implement the simulation, only one vehicle block and one target block need to be built and then copies of these blocks can be used to represent the rest of the vehicles and targets. To simplify simulation model modifications, a vehicle and a target block were implemented and then saved in a SIMULINK block library. This block library was used to create additional instances of the vehicle and target blocks. When one uses a block from a block library, a link from the block to the library is created, so when the library is updated the linked blocks are also updated. Therefore the first vehicle or target block is the *real* block and the rest of the blocks are links to a copy of the *real* blocks in the block library.

Each vehicle model contains the EFS blocks necessary to implement cooperative control of the vehicles. EFS is a collection of software managers that cause the vehicle to

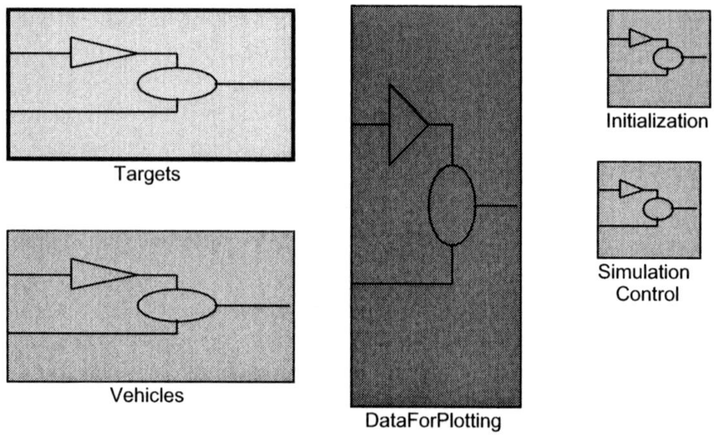


**Figure A.2:** MultiUAV2 top level blocks.

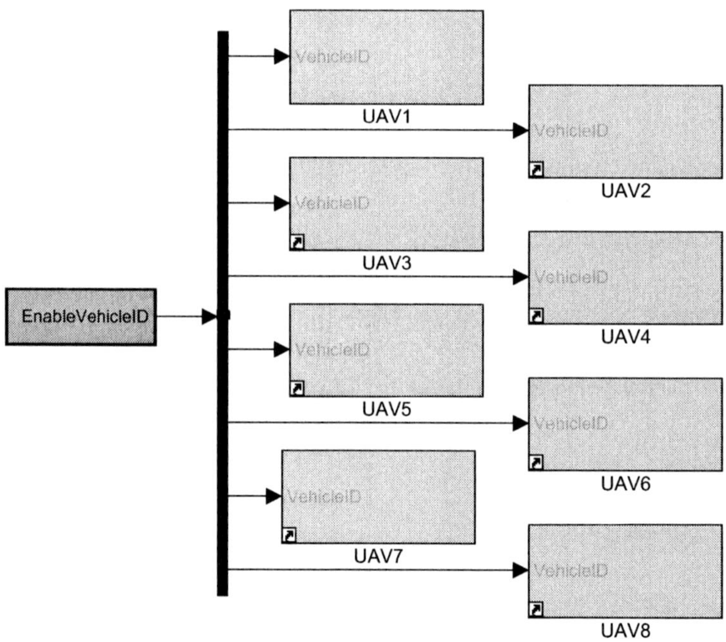


**Figure A.3:** MultiUAV2 vehicle blocks.

perform the desired tasks; see Fig. A.5. The following managers have been implemented:

**Sensor Manager**

This manager is used to perform all of the functions necessary to monitor the sensors and process sensed data. It also contains the sensor process part of the sensor simulation; see section A.3.3.

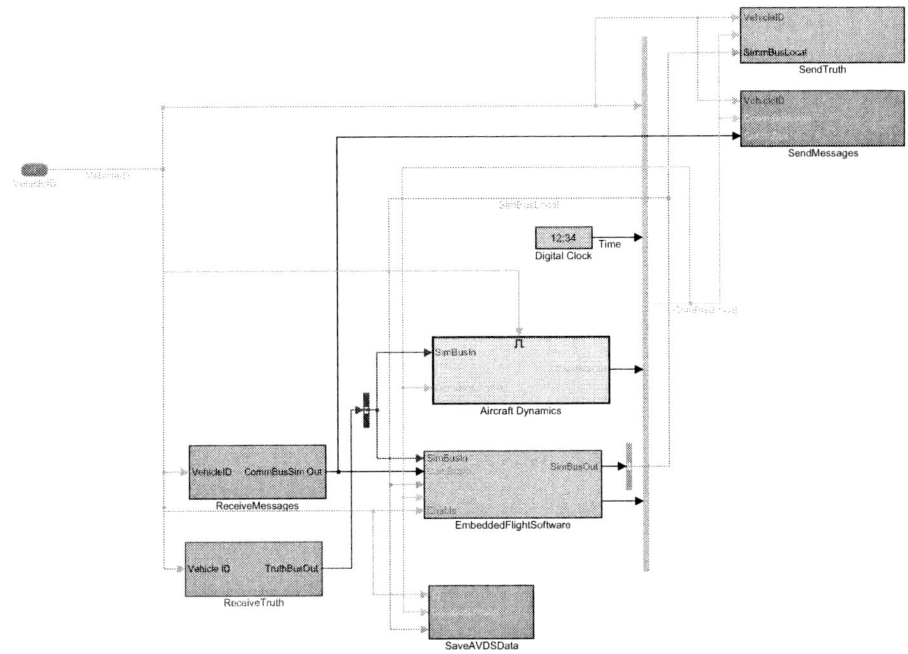


**Figure A.4:** MultiUAV2 blocks that make up a vehicle.

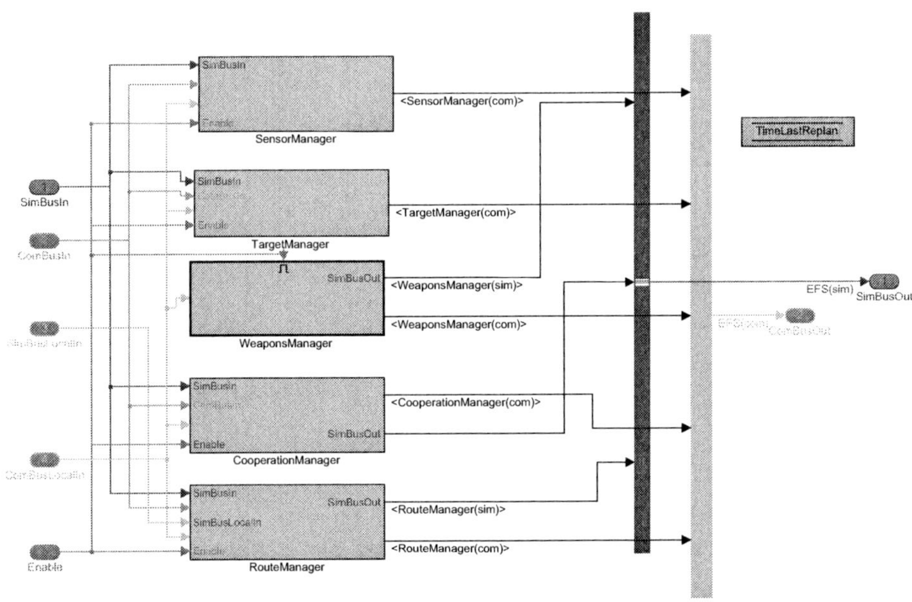


**Figure A.5:** MultiUAV2 managers, embedded flight software.

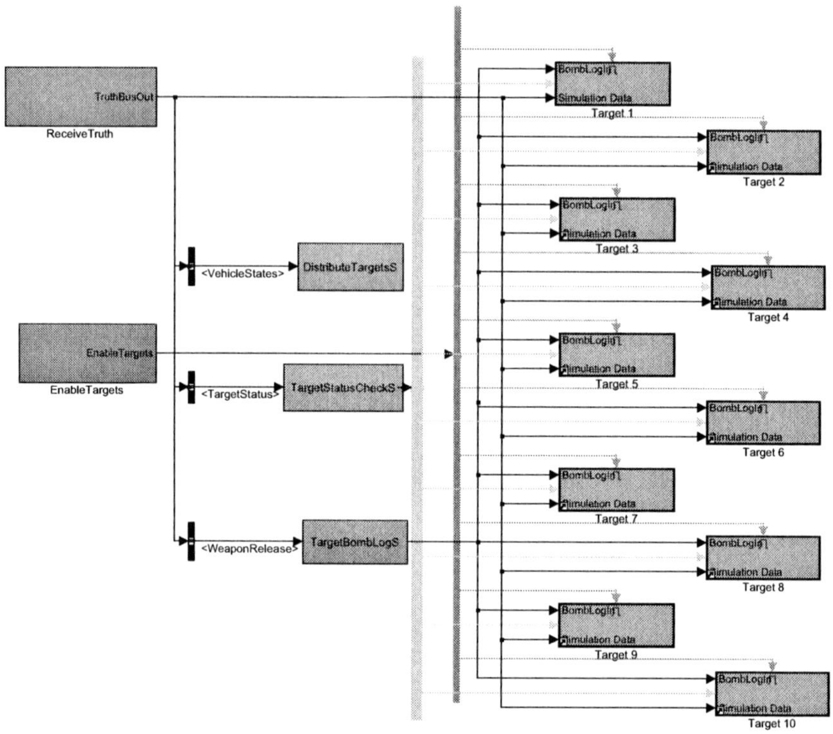


**Figure A.6:** MultiUAV2 target blocks.

**Target Manager**

This manager creates and manages list of known and potential targets. It also manages the target state functions for the vehicle.

**Cooperation Manager**

This manager calculates assignments for the vehicle based on information gathered from all of the vehicles.

**Route Manager**

This manager is used to plan and select the route for the vehicle to fly. Part of the functionality is calculation of the lowest-cost route to all known targets, based on each target's state. The route manager also monitors the status of the vehicle's assigned task.

**Weapons Manager**

The weapons manager selects a weapon and then simulates its deployment.

While it is important to understand the operation of the EFS managers to be able to add new cooperative control algorithms to MultiUAV2, it is also important to understand the other major elements of MultiUAV2. To that end the following sections describe the

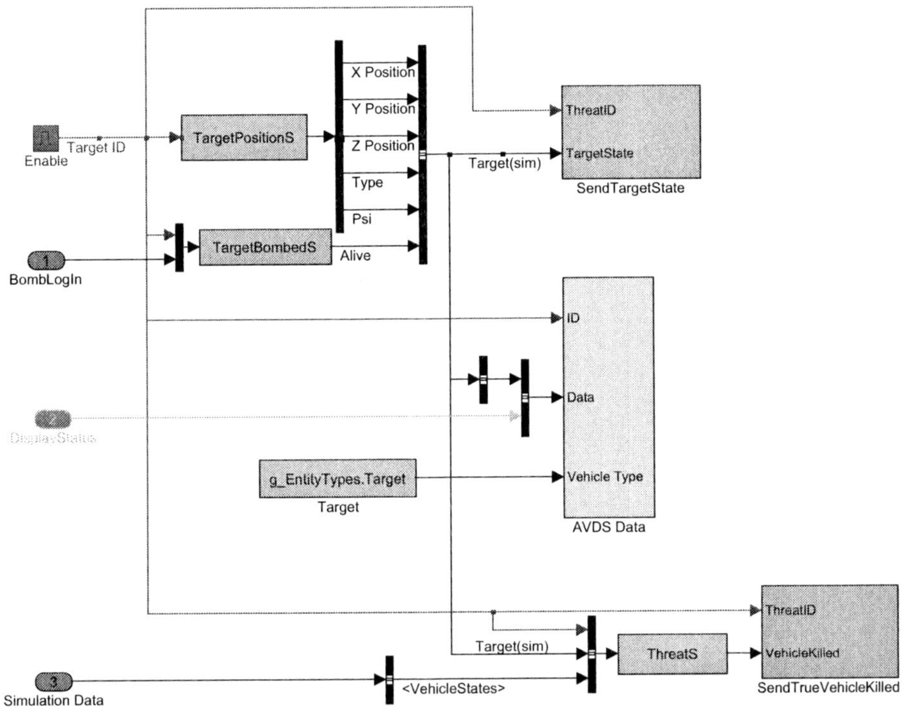


**Figure A.7:** MultiUAV2 blocks that make up a target.

major elements of MultiUAV2, i.e., targets and threats, vehicle dynamics, mission sensors, communications, and cooperative assignment algorithms.

## A.3.1 Target and Threats

Since the targets in MultiUAV2 can be configured to destroy UAVs, they also can act as threats. The nominal targets are fairly simple; i.e., they can calculate where they are, calculate damage from explosions, and calculate if they have shot down a UAV; see Figs. A.6 and A.7. Nominally the targets are stationary, but they can be given trajectories to follow by modifying a MATLAB script file that sets the position of the target.

## A.3.2 Vehicle Dynamics

Vehicle dynamics in MultiUAV2 are generated using a simulation called *Variable Configuration Vehicle Simulation* (VCVS). VCVS is a nonlinear 6DOF vehicle simulation that includes a control system which reconfigures the simulation for new aerodynamic and physical vehicle descriptions. VCVS is written entirely in C++ and can run more that 20 times faster than real time. Vehicle dynamics are based on two configuration files, one containing aerodynamic data and the other physical and control system parameters. The

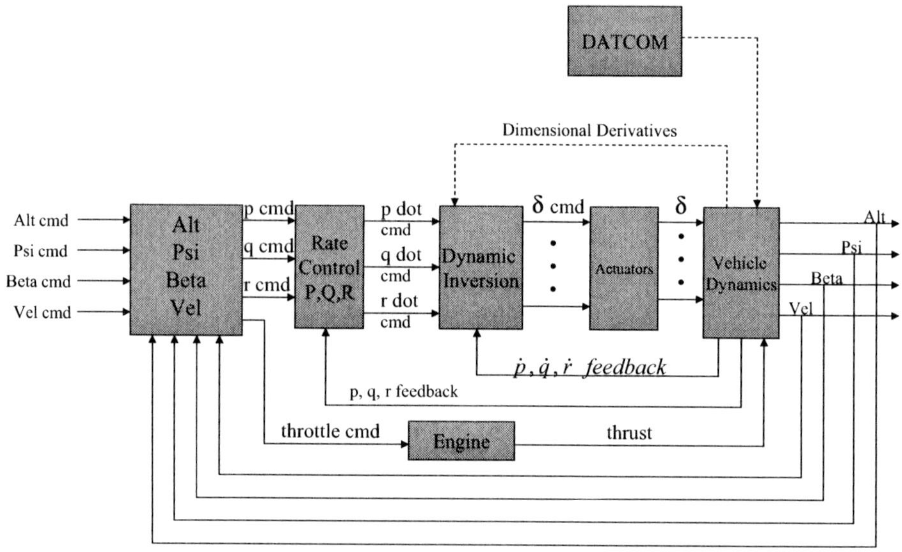


**Figure A.8:** VCVS schematic.

aerodynamic configuration file contains tables of nondimensional forces, moments, and damping derivatives. The vehicle model calculates aerodynamic forces and moments by using the vehicle's state and control deflections as independent variables to look-up values from the aerodynamic tables. During the look-up process, linear interpolation is used for states and deflections not found in the tables. The nondimensional values obtained from the tables are combined with vehicle state data to calculate forces and moments acting on the center of gravity of the vehicle. These forces and moments are combined with external forces and moments, e.g., forces and moments from an engine. The forces and moments are used, along with the physical parameters, to calculate the equations of motion. Included in the model are first-order actuator dynamics, including rate and position limits, and first-order engine dynamics.

VCVS uses a dynamic inversion control system with control allocation as its inner loop control system; see Fig. A.8. A rate control system is wrapped around the inner loop to move from angular acceleration commands to angular rate commands. The outermost loop is an altitude, heading, sideslip, and velocity command control system. Gains for the rate controllers and the outer-loop controllers can be adjusted by changing parameters in the parameter input file. New vehicle dynamics can be introduced by changing the physical and aerodynamic parameters. When new parameters are introduced, the control system uses control allocation to reconfigure for different numbers of control effectors and the dynamic inversion controller compensates for changes in response to control inputs.

Interfaces between VCVS and MATLAB/SIMULINK were constructed to aid in vehicle dynamics development and testing. This made it possible to run VCVS separately from MultiUAV2. VCVS has been used outside of MultiUAV2 to develop algorithms such as formation control, rejoin path control, and surveillance path control. Vehicle models ranging from the size of micro UAVs to large UAVs have been implemented in VCVS.

### A.3.3 Mission Sensors

MultiUAV2 separates the mission sensor simulation into two parts: sensor footprint and sensor process. The sensor footprint is used to determine if a target is visible to the vehicle, and the sensor process is used to produce simulated output from the sensor. This makes it possible to implement the sensor footprint in C++ with the vehicle model, as discussed in section A.3.3. The sensor process in this version of MultiUAV2, simulates an automatic target recognition (ATR) sensor, discussed in section A.3.3.

#### Sensor Footprint

To ensure that the sensor footprint reports every target that it passes over, it must be updated at the same rate as the vehicle dynamics model. Since the vehicle dynamics are updated at a faster rate than the rest of MultiUAV2 it was necessary to implement the sensor footprint in a C++ class, which is executed with the vehicle dynamics. This class contains functions that can calculate rectangular or circular sensor footprints. To check if targets are in the circular footprint, the function compares the difference between each of the target positions and the calculated center of the footprint. If this distance is less than the sensor radius, then the target is in the footprint. The rectangular footprint function transforms the coordinates of the targets into a coordinate system aligned with the rectangular footprint. After transformation the coordinates of each target are compared to the limits of the sensor and any targets inside the limits are reported.

#### ATR Sensor

The major attribute of the ATR sensor simulated in the MultiUAV2 simulation is heading dependence. That is, the quality of target identification of many ATR sensors depends on the direction the target is viewed from. To simulate the functionality of an ATR, heading dependent equations were developed. Given the targets length ($L$), width ($W$), and aspect angle[5] ($\theta$), the single ATR value ($\mathrm{ATR}_s$) is calculated using Eqs. (A.1a) and (A.1b).

$$\mathrm{ATR}_s = \begin{cases} \dfrac{W\arccos(\theta) + L\arcsin(\theta)}{L+W} \times \mathrm{SF} & \text{for } 0 \le \theta \le \dfrac{\pi}{2}, \\ \dfrac{-W\arccos(\theta) + L\arcsin(\theta)}{L+W} \times \mathrm{SF} & \text{for } \dfrac{\pi}{2} < \theta \le \pi, \\ \dfrac{-W\arccos(\theta) - L\arcsin(\theta)}{L+W} \times \mathrm{SF} & \text{for } \pi < \theta \le \dfrac{3\pi}{2}, \\ \dfrac{W\arccos(\theta) - L\arcsin(\theta)}{L+W} \times \mathrm{SF} & \text{for } \dfrac{3\pi}{2} < \theta < 2\pi, \end{cases} \tag{A.1a}$$

$$\mathrm{SF} = 0.8\frac{L+W}{\sqrt{W^2+L^2}}. \tag{A.1b}$$

A representative plot of $\mathrm{ATR}_s$ versus $\theta$ is shown in Fig. A.10. Note that the maximum ATR value is 0.8. In the nominal mission an ATR value greater than or equal to 0.9 is

[5]Angle definitions are shown in Fig. A.9.

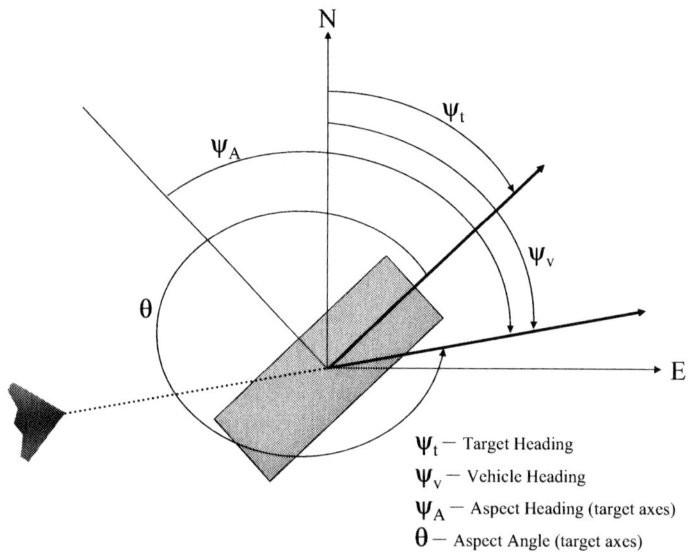


**Figure A.9:** Angle definitions for ATR.

required before a target can be attacked. To increase the ATR value, the value from a single sighting of the target can be combined with another sighting of the target. Given the values for $\text{ATR}_s$ and the respective angles $\theta$, two single ATR values for a target can be combined into one ($\text{ATR}_c$) with the following equations:

$$\text{ATR}_c = (\text{ATR}_1 + \rho \times \text{ATR}_2) - (\text{ATR}_1 \times \rho \times \text{ATR}_2), \tag{A.2a}$$

$$\rho = 1.0 - e^{-0.3|\theta_2 - \theta_1|}. \tag{A.2b}$$

If more than two single ATR values exist for a target, combined ATR values are calculated for all pairwise combinations of the single values. The largest value from the set of combined values is used for that target.

## A.3.4 Intervehicle/Simulation Truth Communications

The MultiUAV2 simulation has two mechanisms for passing messages between objects in the simulation, one for communication messages and one for simulation truth messages. Previous releases of the MultiUAV2 simulation provided vehicle-to-vehicle communication via a signal bus denoted by `CommBus`, while a second aggregated signal bus, `SimBus`, contained the *truth* information for the simulation. The combination of these two data buses represented the complete information state of the simulation. This *perfect* information state was available to all vehicles at every simulation time step. From many perspectives, perfect information access is unacceptable, particularly when considering communication and processing delays. Thus, to incorporate communication and processing delays into MultiUAV2, a new communication framework was introduced; see Fig. A.11. To make

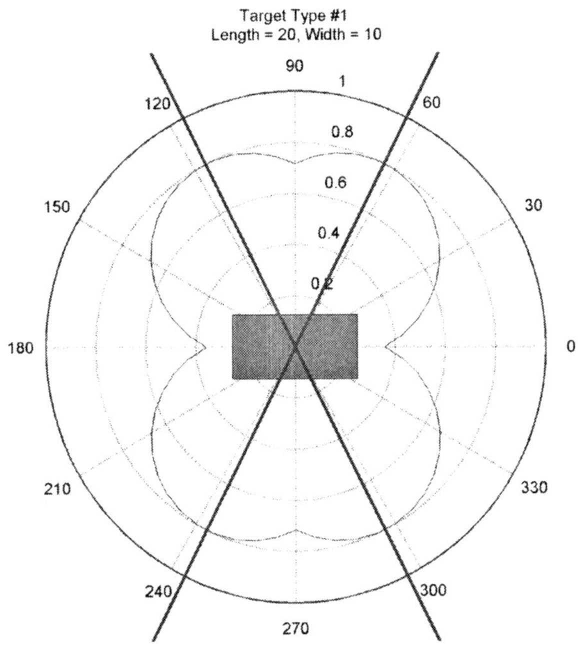


**Figure A.10:** ATR template.

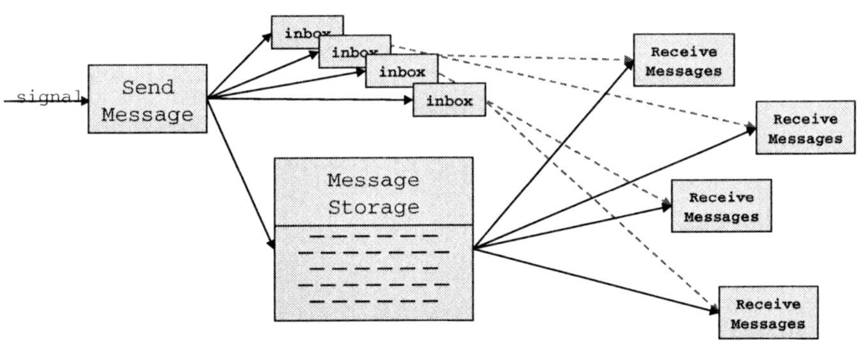


**Figure A.11:** Schematic of the communications framework.

it possible to distribute the MultiUAV2 over many computers, a similar framework was introduced for passing truth information between objects in the simulation.

To provide flexibility for implementation of communication simulations that contain varying levels of detail, a generic message-passing scheme was chosen. In this design, specific message types and their format are defined centrally and made globally available to the various EFS managers as context requires.[6] A message definition must contain a

[6]The message structure discussed here refers to the format dictated by the MultiUAV2 package, rather than to messages related to a specific communication system model.

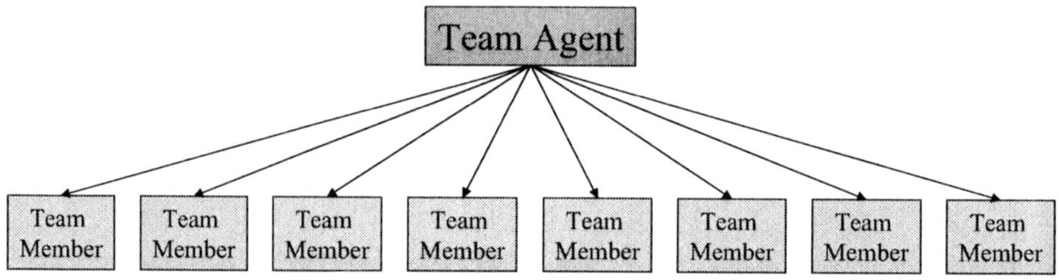


**Figure A.12:** UAV virtual team structure.

unique message identifier, time stamp(s), message layout enumeration, and data field to be written by the EFS manager context. Particular messages may be selected by the EFS managers as output resulting from a computation that must be communicated to other UAVs. Outgoing messages, which include data from each vehicle, are stored centrally and pointers to these messages are distributed to each addressee vehicle's individual input queue. These pointers are composed of the original message header and inform the receiver of the message type, time sent, quality or priority of the message, and which central repository contains the associated message data. A user-defined rule component controls the distribution of incoming messages to all vehicles based on the message headers. For more information on the communications design see [1, 2].

### A.3.5 Cooperative Assignment Algorithms

MultiUAV2 was constructed in parallel with the development of the cooperative assignment algorithms in the following sections. As the algorithms were developed they were implemented and tested using MultiUAV2.

#### Redundant Centralized Implementation

Many of the cooperative control algorithms are implemented in a *redundant centralized implementation* (RCI) manner to control the cooperation of vehicles while they carry out their mission to find, classify, kill, and verify killing the targets in the simulation. RCI consists of vehicles that are formed into a team that contains team members and a team agent, as shown in Fig. A.12. The team agent makes and coordinates team member assignments through the use of a centralized optimal assignment selection algorithm that is based on partial information. The redundant portion of the RCI structure comes about because each team member implements a local copy of the team agent. Because of this, each of the team members calculates assignments for the entire team and then implements the assignment for itself. See Chapter 6, section 6.1, for more details on RCI.

#### Capacitated Transhipment Assignment Problem (Single-Task Tours)

For single-task tours, a capacitated transhipment assignment problem (CTAP) solver is used to calculate the vehicle task assignments. For more information see Chapter 3, section 3.2.

### Iterative Capacitated Transhipment Assignment Problem (Multiple-Task Tours)

For multiple-task tour assignments MultiUAV2 uses an iterative implementation of the CTAP solver. For more information see Chapter 3, section 3.3.

### Iterative Auction (Multiple-Task Tours)

Using the same strategy as the iterative CTAP, the iterative auction builds up multiple task tours of assignments for the vehicles by using a Jacobi auction solver. The auction is used to find an initial set of assignments, freezes the assignment with the shortest estimated time of arrival, and then repeats this process until all possible tasks have been assigned. For more information see the explanation of the iterative CTAP algorithm in section A.3.5

### Relative Benefits (Multiple-Task Tours)

This method requires a relaxation of the optimality requirement but can potentially produce good paths and assignments quickly. One major problem with this and other resource allocation methods is the absence of a good metric to judge its efficacy. Some possible algorithms will return results that are very close to optimum, but none of them has been implemented for this type of problem. The central theme of this algorithm is that multiple assignment tours can be developed by making single-assignment tours and then trading assignments between the UAVs based on the relative benefit of one UAV taking on the assignment of another. For more information on the Relative Benefits algorithm see [3].

### Distributed Iterative Capacitated Transhipment Assignment Problem (Multiple-Task Tours)

The iterative CTAP algorithm was initially implemented in an RCI manner; see section A.3.5. For the distributed CTAP, the original iterative CTAP algorithms were implemented in a distributed manner, i.e., each vehicle calculates benefits for itself to complete the required tasks at each iteration and then sends these benefits to the other vehicles. All the vehicles run the CTAP algorithms and then move on to the next iteration.

### Distributed Iterative Auction (Multiple-Task Tours)

The iterative auction algorithm was initially implemented in an RCI manner; see section A.3.5. For the distributed iterative auction, the original iterative auction algorithms were implemented in a distributed manner, i.e., each vehicle calculates bids for the required tasks at each iteration and sends these bids to the other vehicles for use in an asynchronous distributed auction.

## A.4 Simulation Event Flow Example

During the progress of the simulation, the EFS managers cause the vehicle to react to changes in target states, vehicle tasks, and task assignments. As an example of the information flow between EFS managers during the simulation, when the CapTransShip algorithm is selected the following is a sequence of events that occur when a previously undetected target is discovered:

1. Vehicle Dynamics (sensor footprint) senses target.
2. Based on the vehicle's heading and the target ID, the vehicle's SensorManager calculates a single ATR value.
   - The vehicle sensing the target sends an *ATRSingle* message to all the vehicles. This message contains a time stamp and the target's single ATR value, sensed heading, estimated pose angle, and estimated type.
3. Each vehicle's SensorManager calculates combined ATR values for the target. These values are based on all of the information that the vehicle has obtained about the target.
4. TargetManagers on all vehicles update the target state based on the combined ATR calculated locally.
   - For each vehicle, if any of the targets changes state, the vehicle sends a *TargetStatus* message. This message contains a time stamp and the status of the targets.
5. Based on the change in target status, the RouteManagers, on all of the vehicles, calculate the optimal route, estimated time of arrival (ETA), and cost to perform the required tasks on all of the known targets.
   - Each vehicle sends the *ETACostToSearch* message to all of the other vehicles. This message contains the vehicle's ETA , cost of performing the tasks, and return to search distance for each of the known targets.
6. CooperationManagers on all vehicles calculate optimal assignment for all of the vehicles to perform the next required task on each of the targets based on the supplied costs.
7. RouteManagers on all vehicles implement the route assigned for that vehicle.
8. Tactical Maneuvering Managers functions on all vehicles read assigned waypoints and calculate commands that will cause the autopilot to cause the vehicle to fly to the waypoints.
9. VehicleDynamics reads the autopilot commands and run the vehicle dynamics simulation. This maneuvers the vehicle along the path to perform the assigned task.
10. If the assigned task is attack, the vehicle's WeaponsManager determines where the explosion occurred and sends the *WeaponsRelease* truth message. This message contain a time stamp, weapon ID, weapon type, and weapon aim point.
11. Using the weapon type and aim point, each target calculates damage to it due to the explosion.

12. New replanning cycles are triggered each time a target changes state or if a vehicle fails its assigned task.

## A.5 Summary

The MultiUAV2 simulation has been constructed to be capable of simulating multiple UAVs performing cooperative control missions, while ensuring accessibility to researchers. It is a very flexible simulation that can be modified to perform various missions. Missions are incorporated into MultiUAV2 through modifications to the existing vehicles, targets, and tactics. The vehicle dynamics simulation, VCVS, uses 6DOF equations of motion and nonlinear aerodynamic look-up tables to produce good-fidelity dynamics and can be used separately from MultiUAV2 for other types of research. MultiUAV2 has been used to perform cooperative control research on areas such as task assignment, path planning, and communication effects. MultiUAV2 has been released to the public and has been provided to many government agencies, universities, and companies.

## Bibliography

[1] *MultiUAV2 Simulation User's Manual.* AFRL/VACA, Wright-Patterson AFB, OH, 2005.

[2] J. W. Mitchell, S. J. Rasmussen, and A. G. Sparks. Communication requirements in the cooperative control of wide area search munitions using iterative network flow. In *Theory and Algorithms for Cooperative Systems, Series on Computer and Operations Research*, Vol. 4, D. Grundel, R. Murphy, and P. Paradalos, eds., World Scientific, Singapore, 2004.

[3] S. J. Rasmussen, C. J. Schumacher, and P. R. Chandler. Investigation of single vs. multiple task tour assignments for UAV cooperative control. In *Proceedings of the 2002 AIAA Guidance, Navigation, and Control Conference*, Monterey, CA, 2002.

[4] AVDS. RasSimTech Ltd. http://www.RasSimTech.com.

Appendix B

# Path Planning for UAVs

*Nicola Ceccarelli and Steven Rasmussen*

*No matter how far you have gone on the wrong road, turn back.*
—Turkish proverb

The requirements of scenarios where cooperative multiple UAV systems face multitarget multitask missions lead to a growing importance for the role of the path planner module. Indeed "off-the-shelf" UAV solutions are usually equipped with reliable autopilot units providing fairly complete and practically inaccessible guidance navigation and control (GNC) modules. An abstraction of the concept is diagrammed in Fig. B.1, where the UAV system is envisioned as equipped with a GNC unit that steers the vehicle over the path planned, based on a specified *Mission*.

The responsibility of the final user is that of designing the *Path Planner* module and providing criteria and techniques to evaluate whether the mission requirements are met, based on the data flow coming from the vehicle sensors.

The GNC module gives the vehicle the ability to execute unmanned flight over a predefined path. Although its design is by far the most relevant aspect for autonomous vehicles, any attempt of a minimally rigorous treatment is well beyond the aim of this book. The reader is referred to [7], and references therein, and to [1, 10, 11] for examples of commercial products available on the autopilot market.

Path planning is discussed in section B.1. Path guidance autopilots, discussed in section B.2, determine how the UAV follows such a given path. UAV modeling for path planning are discussed in section B.3. L. E. Dubins laid out a method to find a kinematic solution to this problem [4]. This is discussed in section B.4, along with guidelines for a possible implementation. Finally, a possible extension to the relevant path planning in wind is proposed in section B.5.

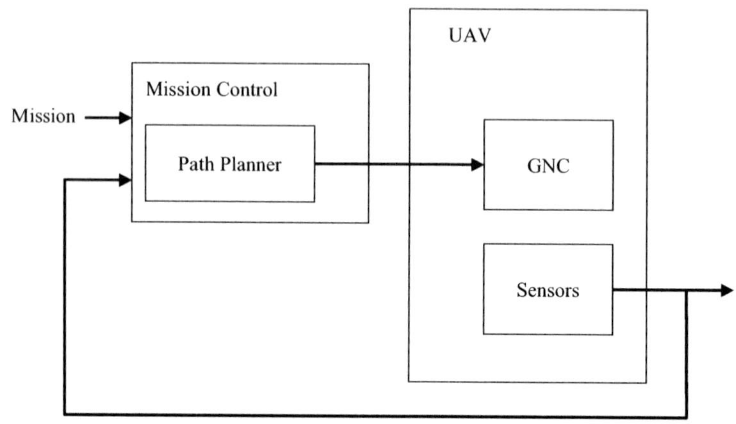


**Figure B.1:** Mission control loop.

## B.1 Path Planning

Any practical cooperative control algorithm for UAVs must include the functionality necessary to calculate the paths required by the UAVs to navigate their assigned tasks. Depending on the requirements of the algorithms, and the UAVs, these paths can range from simple point-to-point paths to smooth paths that account for vehicle dynamics. Once a path is generated, the cost of the assignment can be calculated and used in the vehicle assignment process and, if selected, the path can be used by the vehicle to perform the task. There are many different variations on the path planning problem that can introduce different types of constraints. For instance, the vehicles may be required to operate in a region where there are obstacles or threats that must be avoided at all costs. In other scenarios a vehicle's path may be coupled to a team member's, making it necessary to coordinate their movements. To simplify this discussion, the path planning variation that is used in this appendix is one where each vehicle's path is independent of the others and there are no obstacles. Furthermore, we will limit our path planning discussion to paths required by fixed-wing UAVs. A detailed review on classical path planning approaches and formulations can be found in [6].

To construct an efficient, responsive, and cooperative control system, it is desirable to produce paths that minimize parameters such as the path length and the time required to traverse the path. In particular we tackle the problem of generating a path between assigned initial and final poses. Since the scenarios of interest do not include path coupling and obstacles, it follows, trivially, that the shortest path is a line between the starting point and the end point. However this path is usable only if the state and dynamics of the UAV permit. That is, since the UAVs cannot initiate a turn instantly and have a minimum turn radius, the planned path should incorporate changes in direction that do not violate these constraints. Moreover, the paths of these UAVs can be further constrained by the GNC module incorporated in the autopilot.

## B.2 Path Guidance

One assumption implied in the previous discussion is that the UAV's control system will be able to cause the UAV to follow the plan. This means that during the path design, one should take into account the vehicle dynamics and the type of guidance system to cause the UAVs to follow the planned paths. Three examples of basic path following schemes are presented here—*waypoint following*, *cross-track following*, and *path following*.

### B.2.1 Waypoint Following

This type of guidance scheme uses only the position of the waypoints to "navigate" the path. The autopilot commands the UAV to steer in order to line up its course of travel with the direction of the next waypoint, $P_{i+1}$; see Fig. B.2(a). That is, the autopilot causes the vehicle to steer in order to minimize the error, $\tau_\psi$, between the UAV's heading and the azimuthal direction of the line of sight of the waypoint relative to the UAV. Moreover, when the UAV is within a specified distance from the waypoint, depicted as a dashed disc in Fig. B.2(a), the autopilot commands the vehicle to change heading to line up with the next waypoint.

Fig. B.2(b) depicts a simulation where a vehicle navigates the path described as a preassigned sequence of waypoints using a *waypoint following* guidance scheme. As shown in the figure, the trajectory of the vehicle, the dashed line, loosely follows the intended path, indicated by the solid line. As the vehicle comes within $\rho_{wp}$ of each waypoint it steers toward the next. This can cause large angular errors between the UAV's course and the intended path. Based on the requirements of the mission, this may be acceptable. If it is important to follow the intended path more closely, extra waypoints can be added to the planned path. That is, waypoints can be added if the vehicle can accommodate them. For instance, if it is important to fly over an intended target from a preassigned direction, extra waypoints can be added before and after the target to force the UAV to fly over the target with the correct course direction.

Waypoint following is relatively simple and efficient to implement so it can be used on UAVs with little processing ability. If the distance to travel between waypoints is relatively far, then this type of autopilot will work well. However, if it is important for the vehicle to follow a particular path or if it is important to arrive at a waypoint from a particular heading, then the waypoint following autopilot may not be suitable.

### B.2.2 Cross-Track Following

A guidance system based on cross-track following causes the vehicle to navigate the path made up of the line segments or tracks, joining each consecutive pair in the preassigned waypoint sequence. The concept is illustrated in Fig. B.3(a). The current track is the line segment between the last flown waypoint and the next waypoint. The autopilot steers the vehicle to minimize the cross-track error distance $\tau_T$.

Cross-track following autopilots are slightly more difficult to implement than waypoint following but are still simple enough that they can be employed on relatively simple processors. Fig. B.3(b) shows a simulation for a vehicle equipped with a cross-track following guidance system. The figure shows that the UAV, the dashed line, follows the intended path, the solid line, fairly well, making excursions only when the turn radius of the path

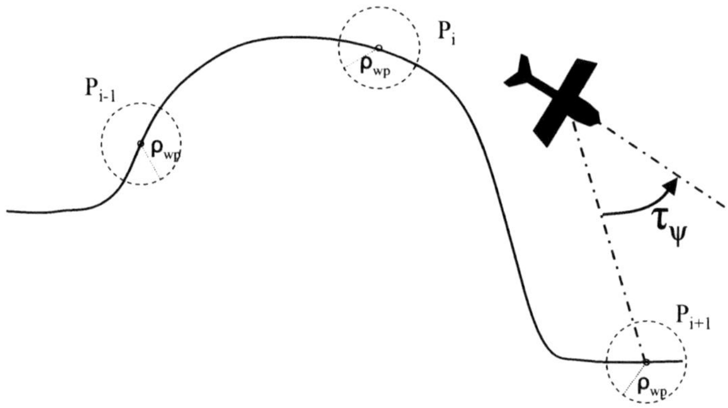


(a) guidance

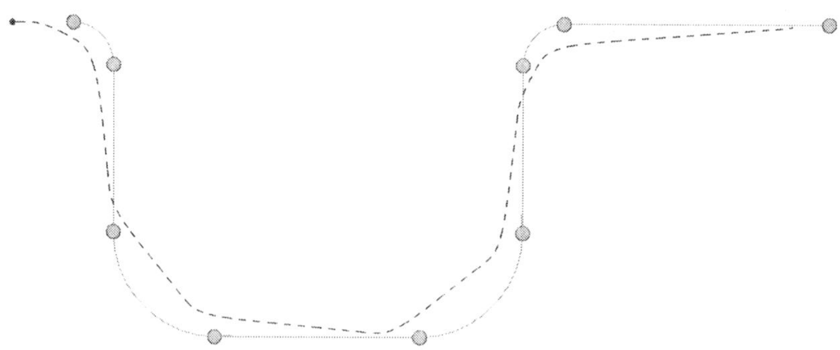

(b) UAV trajectory

**Figure B.2:** Waypoint following guidance.

is significantly larger than the minimum turn radius of the vehicle. Again, the plan can be modified by adding waypoints to cause the UAV to follow the path more closely, if the UAV can accommodate them. For instance, on the corners of Fig. B.3(b) where the turn radius is large, extra waypoints can be added on the turn to "pull" the UAV along the radius of the turn.

### B.2.3 Path Following

Path following autopilots regulate the vehicle to the intended path by minimizing the distance between the path and the UAV position, $\tau_P$; see Fig. B.4(a). A path following autopilot requires the description of a smooth continuous flyable path. One method of providing such an abstraction is through the use of waypoints that include entries describing the path. Table B.1 lists possible entries of a waypoint that may be suitable for the description of continuous smooth paths.

By adding the coordinates of a turn center and turn direction to the waypoint definition, these waypoints can describe circular arc paths as well as linear segments. To ensure that the

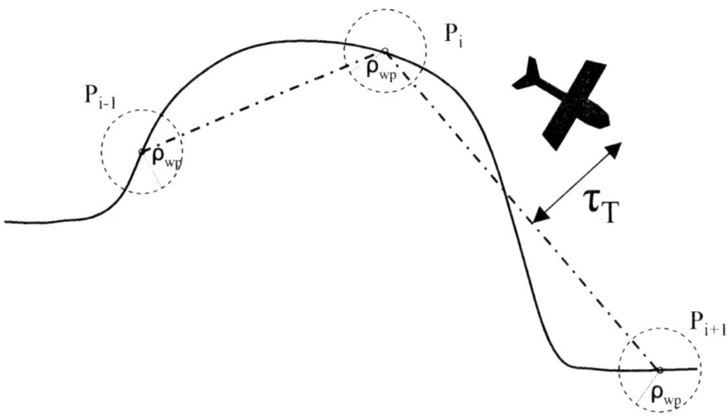


(a) guidance

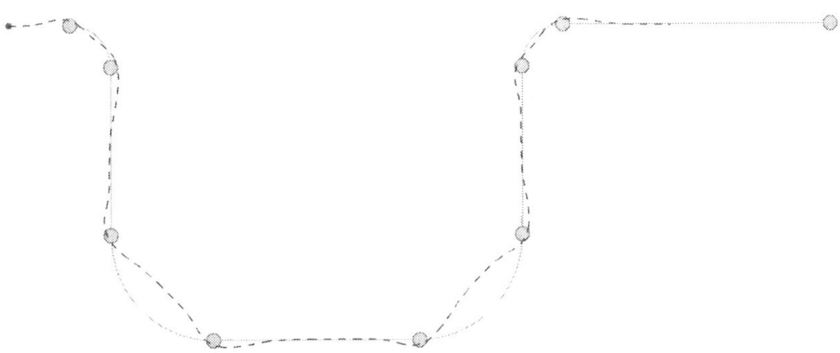

(b) UAV trajectory

**Figure B.3:** Cross-track following guidance.

path is flyable, the designer must constrain the turn radius to be greater than some preassigned value $\rho_{min}$, which is based on the vehicle dynamics. Moreover, to avoid undesirable sharp change of course, the planner needs to ensure that the projection of the path on the two-dimensional plane, i.e., flat earth, is a curve of at least class $C^1$ (each component of the associated map has continuous first derivative). In other words, no jumps of the heading are allowed. The planned path in Fig. B.4(b) was designed and specified in this manner.

Path following guidance systems can be much more complicated to implement than the other two. However, they can provide much more control over the actual path that the UAV travels; see Fig. B.4(b). As depicted in the figure, the UAV, indicated by the dashed line, follows the path well, but there are points, mainly noticeable in tight turns, where the vehicle misses the path. This happens because the path was constructed under the assumption that the vehicle could instantaneously attain a commanded turn rate, which is not true for actual vehicles.

Any path planning algorithm must account for the dynamics of the vehicle and the path guidance system. This can be as simple as using a minimum waypoint separation or

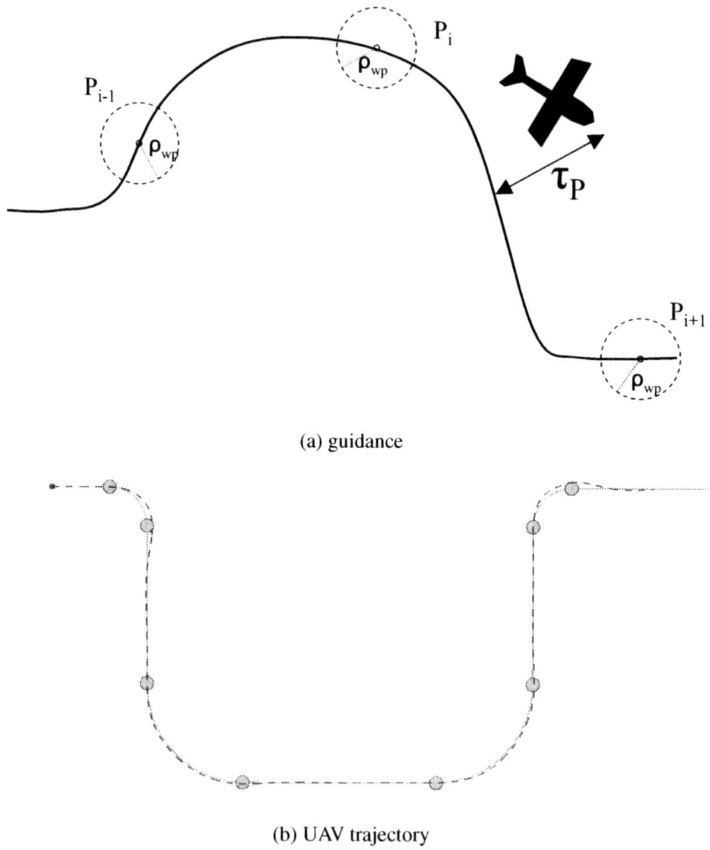


**Figure B.4:** Cross-track following guidance.

**Table B.1:** Waypoint entries.

| | |
|---|---|
| **Position:** | Latitude<br>Longitude<br>Altitude |
| **Turn Center:** | Latitude<br>Longitude<br>Direction |

as involved as planning a continuous flyable path. Since the trajectory that the UAV flies is affected by both the planned path and the path guidance autopilot, it is important to select an autopilot which can meet the preassigned mission requirements. Table B.2 summarizes some of the attributes of the three path guidance systems discussed in this section.

**Table B.2:** Comparison of attributes of guidance following autopilots.

| Attribute | Waypoint | Cross-Track | Trajectory |
|---|---|---|---|
| Ease of implementation | ⋆⋆⋆ | ⋆⋆ | ⋆ |
| Fly over a point | ⋆⋆⋆ | ⋆⋆⋆ | ⋆⋆⋆ |
| Point with prescribed heading | ⋆ | ⋆⋆⋆ | ⋆⋆⋆ |
| Follow a path | ⋆ | ⋆⋆ | ⋆⋆⋆ |
| Predict path timing | ⋆ | ⋆⋆⋆ | ⋆⋆⋆ |

## B.3 Vehicle Model

Many UAVs can be modeled as rigid structures with six-degrees-of-freedom equations of motion and nonlinear aerodynamic forcing functions. These are usually complex and fairly complete models that are exploited mainly in the development of simulation tools. However, a path planning strategy based on a detailed dynamic model may result in an unnecessarily complicated analysis. Indeed, the autopilot guidance unit, rather than requiring the user to specify the vehicle surface control inputs, allows the user to specify a path to be followed. In this respect restrictions and simplifications are requisite in order to develop a path planner that takes into account the performances and characteristics of the vehicle. In particular for the purpose of path planning we assume constant velocity, constant altitude, and minimum turn radius. While the first two are derived from the mission requirements on cruise speed and altitude, the full UAV model can be used to calculate the minimum turn radius that depends on the aerodynamic and structural force limits of the UAV.

### B.3.1 Nonholonomic Kinematic Constraints

Since fixed-wing UAVs have the motion velocity vector constrained along the vehicle body axis, they fall into the class of vehicles with nonholonomic constraints. Nonholonomic constraints are those constraints that cannot be integrated to yield constraints solely based on the position variables of the vehicle equations of motion [9]. A fixed-wing UAV's motion is not just nonholonomic, it also has to maintain a minimum forward speed to fly, hence resulting in a minimum turn radius. This means that the vehicle's minimum turn radius must be included in path planning explicitly or implicitly. For example, if the path's start and end points are far apart when compared to the turning radius of the UAV, then the minimum turn radius constraint could be satisfied, implicitly, by connecting the two points with a straight line. On the other hand, if the points are close together, when compared to the minimum turn radius, then it may be necessary to generate curved paths that connect the points, thereby satisfying the turn constraint explicitly.

### B.3.2 Flyable Paths

To efficiently move a UAV from one point to another, in a predictable manner, the UAV must be able to follow the planned path. If the path is a straight line, this may not be a concern, but if there are many turns in the path, some of the assumption used in construction of the plan can be severely tested. A simplistic UAV kinematic model, mainly characterized by constraints on minimum allowable turn radius, may neglect vehicle dynamics such as those involved in initiating and recovering from turns. Therefore, if the path is based on the minimum turn radius alone, then the actual vehicle dynamics can prevent the vehicle from closely following the planned path. Based on vehicle and mission requirements, the allowed turn radius can be consequently adjusted. The consequences of the vehicle not closely following the planned path may vary depending on the mission. For instance, if it is important to cause the vehicle to fly over a region with no constraints on the time over the area of interest or the time of flight, then following the path loosely may not be of concern. On the other hand, if it is important to fly over a point from a prescribed angle or at a prescribed time, then it may be critical for the vehicle to closely follow the planned path.

## B.4 Optimal Paths

Since the purpose of this appendix is to investigate possible path planning strategies and in agreement with the discussion in the previous sections, the vehicle kinematic model is reduced to a simple unicycle model with constant forward velocity. This removes the altitude as a degree of freedom, which is commonly a mission requirement; moreover, a bounded turn rate is assumed in order to take into account the constraint on the minimum turn radius. With these assumption the kinematic equations are

$$\dot{x}_V(t) = V_a \cos(\theta_V(t)), \tag{B.1}$$

$$\dot{y}_V(t) = V_a \sin(\theta_V(t)), \tag{B.2}$$

$$\dot{\theta}_V(t) = u(t), \tag{B.3}$$

$$|u(t)| \leq \frac{V_a}{\rho_{min}} [\text{rad/s}]. \tag{B.4}$$

The objective is to generate a path for the vehicle to follow, from a given pose $[x_V, y_V, \theta_V]$ that ends at a given pose $[x_F, y_F, \theta_F]$. The path must minimize either path length or the flight time. We further assume a negligible wind condition, that is, the case of a UAV flying at cruise speed significantly greater than the actual wind speed. With these assumptions, finding the minimum length path is equivalent to finding the minimum time path. We will tackle with the wind case for low cruise speed UAV in section B.5.

### B.4.1 Minimum-Length Path

The problem of finding the shortest path to steer a vehicle, subjected to the kinematic of Eqs. (B.1)–(B.4), from an initial pose $[x_V, y_V, \theta_V]$ to a final pose $[x_F, y_F, \theta_F]$, was first solved in 1957 by Dubins [4]. In particular it can be shown [2, 4] that the minimum-length path is given by one of two forms of maneuvers, Turn-Turn-Turn or Turn-Straight-Turn, where *Turn* means a turn executed with radius equal to $\rho_{min}$, i.e., $u = \frac{V_a}{\rho_{min}}$ in Eq. (B.3). As shown in the following table, there are six of these possible maneuvers:

| |
|---|
| Right-Straight-Right |
| Right-Straight-Left |
| Left-Straight-Left |
| Left-Straight-Right |
| Right-Left-Right |
| Left-Right-Left |

Figs. B.5(a) and B.5(b) show optimal paths that use Right-Left-Right and Right-Straight-Left maneuvers to move from one pose to another. Guidelines for the implementation of a minimum-length path algorithm are provided in the following section.

## B.4.2 Dubins Algorithm Implementation

To generate paths for the MultiUAV2 simulation (see Appendix A), a function was constructed based on the Dubins set of maneuvers mentioned above. The function, `Minimum Distance`, is used to calculate the minimum time route from the vehicle's current pose to an assigned target. The required input parameters are (see Fig. B.6)

- vehicle pose $[x_V, y_V, \theta_V]$,
- target position $[x_T, y_T]$,
- desired approaching azimuthal direction $\theta_F$,
- desired target approaching distance ("stand-off") $\rho_{off}$,
- commanded turn radius.

The function structure is briefly described in the following steps (see Fig. B.7):

1. Calculate the centers, $O_{v1}$ and $O_{v2}$, of the turn circles that are connected to the vehicle using the vehicle's pose and minimum turn radius.
2. Calculate the point $S_{off}$, based on the target position $T$, desired azimuthal approaching direction $\theta_F$, and stand-off distance $\rho_{off}$.
3. Calculate the centers of the turn circles $O_{t1}$ and $O_{t2}$, of radius $\rho_{min}$, tangent at $S_{off}$ and approaching the target from desired direction $\theta_F$.
4. Determine the appropriate turn directions for all $O_{v1}$, $O_{v2}$, $O_{t1}$, and $O_{t2}$, to be consistent with the path initial and final poses.
5. Consider now each possible pair of turn circles $\{O_v, O_t\}$, where $O_t \in \{O_{t_1}, O_{t_2}\}$ and $O_v \in \{O_{v_1}, O_{v_2}\}$. We denote with $c_t$ and $c_v$ the center of $O_t$ and $O_v$, respectively. Based on the relative position between the two, different possible tangent connections of arcs and segments are taken into account; see Fig. B.8.
6. Exploit the specified direction of rotation for $O_v$ and $O_t$ to select arcs and segments based on their suitability to originate a feasible connecting path between the vehicle pose and the desired pose $S_{off}$.
7. Select the shortest, among all the solution paths, originated by each pair $\{O_v, O_t\}$.

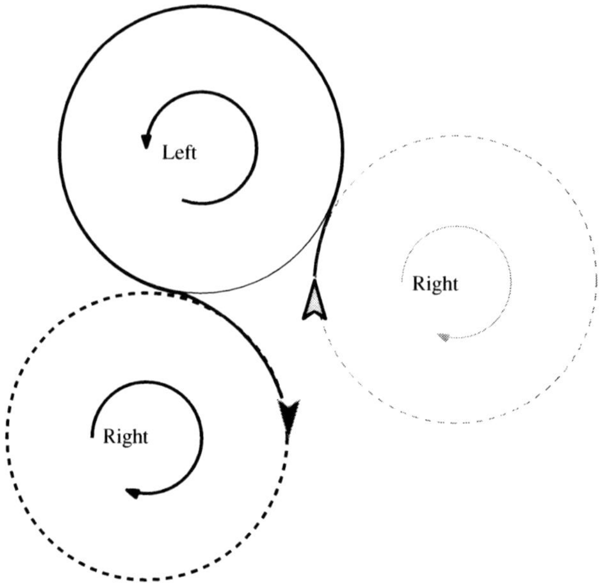


(a) Turn/Turn/Turn

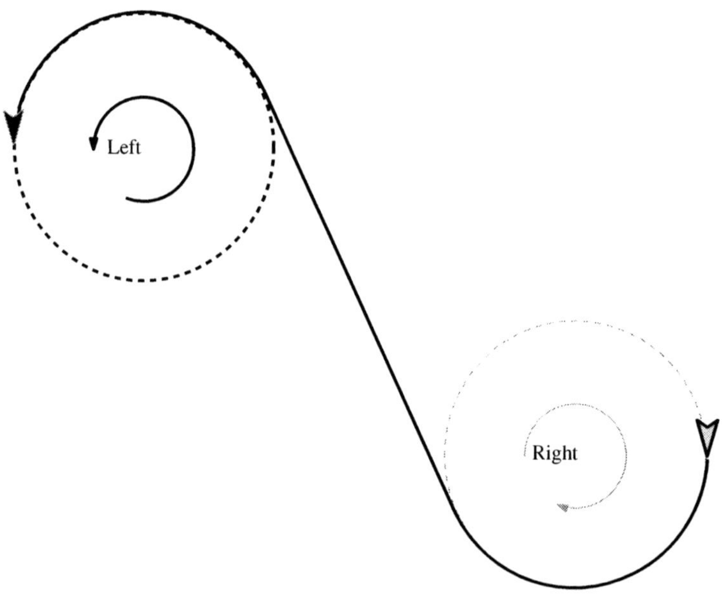


(b) Turn/Straight/Turn

**Figure B.5:** Examples of possible maneuvers.

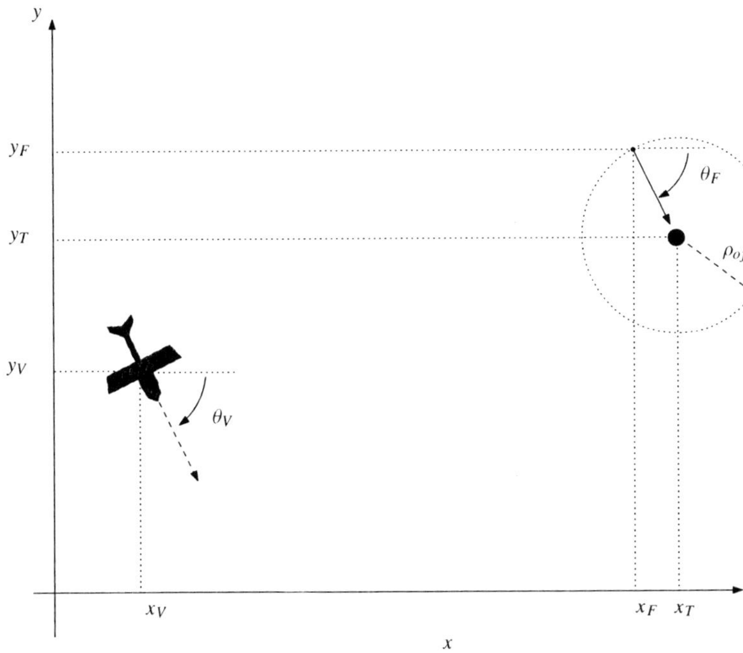


**Figure B.6:** Minimum-length path problem.

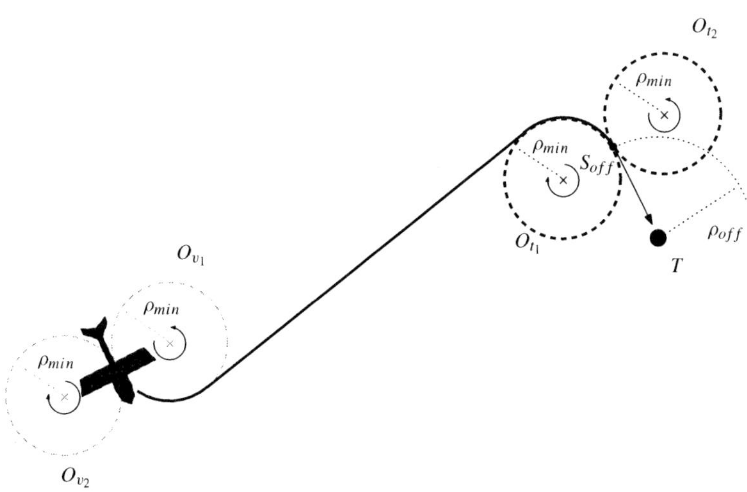


**Figure B.7:** Minimum-length path solution.

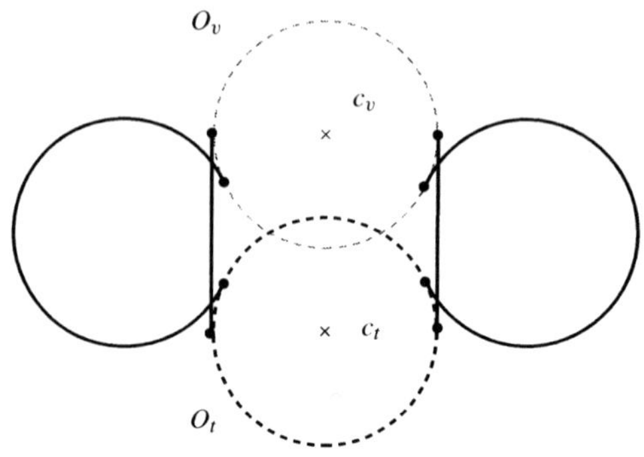


(a) $\|c_v - c_t\| \leq 2\rho_{min}$

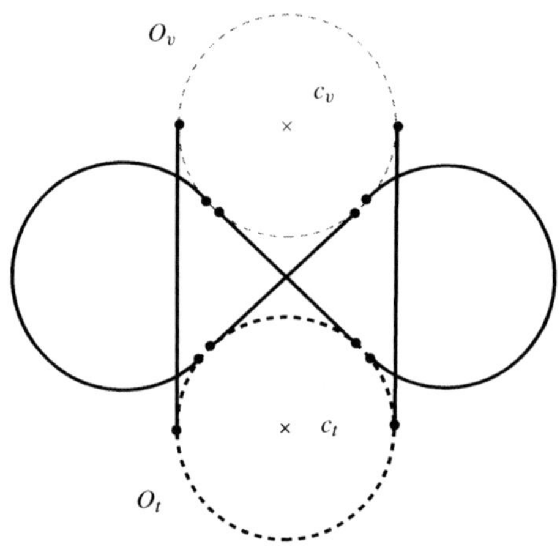


(b) $2\rho_{min} \leq \|c_v - c_t\| \leq 4\rho_{min}$

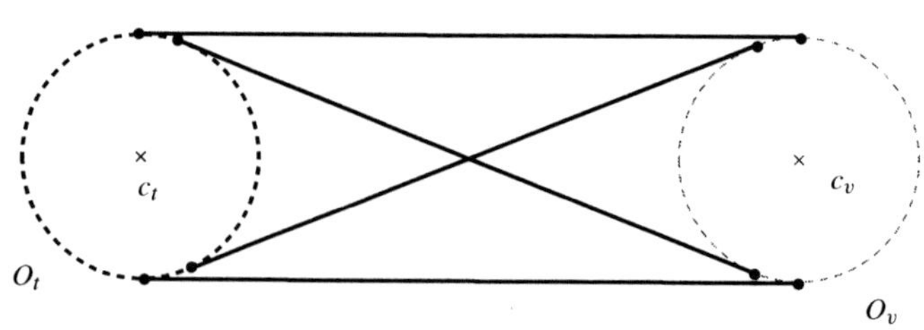


(c) $\|c_v - c_t\| \geq 4\rho_{min}$

**Figure B.8:** Possible tangent connections between two circles.

## B.5 Waypoints Path Planning in Wind

This section deals with scenarios in which the vehicle is subjected to a known constant wind that significantly affects the vehicle ground speed along the path. The approach proposed is based on the practical assumption that control can be exercised only through the selection of waypoints [3].

### B.5.1 Waypoints Path Planning

As mentioned in section B.2, the guidance system, part of the vehicle autopilot, heavily effects the actual path followed by the UAV. For example, in Fig. B.9 the vehicle, an Applied Research Associates' Nighthawk Micro Aerial Vehicle (MAV) [5] was guided by a Kestrel autopilot [10] with cross-track path following implemented; see section B.2.2. The GPS ground track from a flight test is plotted (dashed line) versus the optimal Dubins path (solid line), generated as described in section B.4, and the associated waypoints sequence, resulting from curve quantization. The actual path turns out to be closer to the segment sequence (dotted line) than to the optimal Dubins path.

Each time the MAV reaches a waypoint, it begins to follow the next line segment, and it may exhibit some transient behavior characterized by a damped oscillation that decays to the desired steady state flight along the new line. In some cases, the line segment may not be sufficiently long for the MAV to reach steady state, or the change in desired heading is too drastic, and the MAV may miss the next waypoint. Rather than relying on this transient response, we propose to choose waypoints in such a way that upon reaching each waypoint, the MAV has reached steady state flight along the previous line segment. This is done by placing an upper bound on the heading difference between sequential line segments and a lower bound on line segment length.

Given an initial and a final pair of waypoints,

$$X_{\text{init}} = [X_{\text{init}_1}, X_{\text{init}_2}] \text{ and } X_{\text{final}} = [X_{\text{final}_1}, X_{\text{final}_2}],$$

the problem is that of finding a number of steps $N \in \mathbb{N}$ and an input sequence $\mathbf{u} = [u(1), \dots, u(N-1)]$ such that

$$\begin{aligned} X(k+1) &= X(k) + u(k) \qquad (k = 1, \dots, N-1), \\ &\text{subject to} \\ X(0) &= X_{\text{init}_1}, \\ X(1) &= X_{\text{init}_2}, \\ X(N-1) &= X_{\text{final}_1}, \\ X(N) &= X_{\text{final}_2}. \end{aligned} \tag{B.5}$$

Once a path has been found, the autopilot inner control loop will be responsible for flying it, based on the sequence $X(k)$ of waypoints. Hence, we introduce additional constraints so that the resulting path may be easily followed by the autopilot. Specifically, we wish to minimize the transient response to direction change in terms of settling time and peak error between the actual trajectory and the desired path. In Fig. B.10, possible

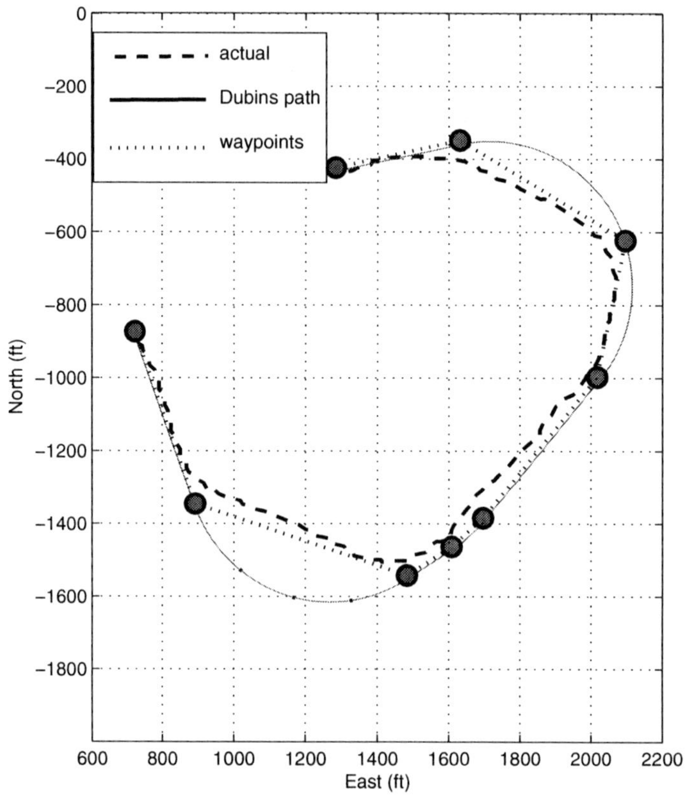


**Figure B.9:** Flight test ground track for MAV following waypoints drawn from the Dubins optimal path.

transient responses of the MAV for different changes of direction are depicted. In particular, let us define

$$\theta(k) = (\measuredangle u(k) - \measuredangle u(k-1)) \bmod \pi \tag{B.6}$$

as the change in direction performed at the step $k$. We introduce the angular constraints

$$|\theta(k)| \leq \alpha \qquad (k = 2, \ldots, N-1), \tag{B.7}$$

where $\alpha$ is the maximum magnitude allowed of the change in direction at each step of the waypoint path. This constraint is meant to keep the peak error within a reasonable level. Furthermore, we introduce a requirement on the minimum length of each step. This is designed to allow the MAV to reach steady-level flight before the next change in direction occurs:

$$\|u(k)\| \geq L \qquad (k = 1, \ldots, N-1). \tag{B.8}$$

In many ways, the building blocks of our path planner are line segments, not waypoints.

We search for a minimum-length path, feasible with respect to system (B.5) and for which constraints (B.7) and (B.8) hold. Taking inspiration from work on the Dubins vehicle

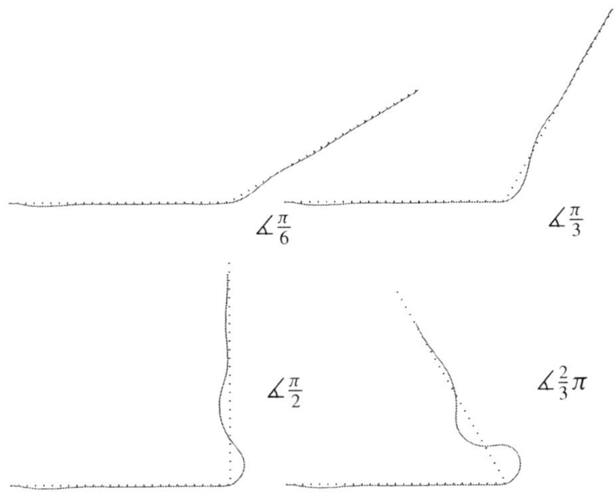


**Figure B.10:** Transient response due to direction changes.

[2, 4], we restrict the search to the following class of control sequences:

$$\bar{\mathcal{U}} \in [u_{\text{init}}\ \mathcal{T}_{\text{init}}\ \mathcal{C}\ \mathcal{T}_{\text{final}}\ u_{\text{final}}], \tag{B.9}$$

where $\mathcal{T}_{\text{init}}, \mathcal{T}_{\text{final}} \in \mathcal{T}$ represent input sequences performing a right or left turn at the maximum rotational rate, i.e., constraint (B.7) is saturated; $\mathcal{C}$ represents the connecting step between the initial and final turn. In particular we define

$$u_{\text{init}} = X_{\text{init}_2} - X_{\text{init}_1}, \tag{B.10}$$

$$u_{\text{final}} = X_{\text{final}_2} - X_{\text{final}_1}, \tag{B.11}$$

$$\mathcal{T} \triangleq \Big\{[u_{\mathcal{T}} \ldots u_{\mathcal{T}}] \in \mathbb{R}^{2\times N_{\mathcal{T}}} : N_{\mathcal{T}} \in \mathbb{N}, \quad u_{\mathcal{T}} = L \cdot \begin{bmatrix} \cos(\alpha_{\mathcal{T}}) \\ \sin(\alpha_{\mathcal{T}}) \end{bmatrix}, \alpha_{\mathcal{T}} \in \{\alpha, -\alpha\}\Big\}, \tag{B.12}$$

$$\mathcal{C} \triangleq \big\{u \in \mathbb{R}^2 : \|u\| \geq L \vee \|u\| = 0\big\}. \tag{B.13}$$

Although the issue of the existence of such a path for any two pairs of waypoints has been neglected, we should report that counterexamples can be generated by choosing arbitrarily close initial and final waypoint pairs. Simulations have shown this event to be rare, and when it does occur, it can simply be overcame by choosing a slightly wider $\alpha_{\mathcal{T}}$, feasible with respect to (w.r.t.) constraint (B.7), for one of the two initial and final turn $\mathcal{T}_{\text{init}}, \mathcal{T}_{\text{final}}$.

### B.5.2 Path Planning in Wind

The formulation presented above does not consider the heavy effect of a wind field on the flight performance of the MAV. Constraints (B.7) and (B.8) correspond to performance requirements in a zero wind field, based on the transient response of the MAV controlled by the autopilot. It is possible to transform the path planning problem in the presence of a known constant wind into one with zero wind, by considering a moving target, with velocity opposite to the actual wind velocity [8]. To maintain consistency with the original path planning problem, initial and final constraints on (B.5) must be generated as follows:

$$\tilde{X}(0) = \tilde{X}_{\text{init}_1}, \tag{B.14}$$

$$\tilde{X}(1) = \tilde{X}_{\text{init}_2} = X_{\text{init}_2} - T_2 \cdot V_W, \tag{B.15}$$

$$\tilde{X}(N-1) = \tilde{X}_{\text{final}_1} = X_{\text{final}_1} - T_{N-1} \cdot V_W, \tag{B.16}$$

$$\tilde{X}(N) = \tilde{X}_{\text{final}_2} = X_{\text{final}_2} - T_N \cdot V_W, \tag{B.17}$$

where $V_W$ is the wind velocity vector and $T_k$ is the time required to arrive at the $k$th waypoint. It should be noted that the value of $T_k$ depends on the particular path considered and its computation requires the solution of a quadratic equation.

Now we look for an input sequence $\tilde{\mathbf{u}}^* \in \tilde{\mathcal{U}}$ that gives the minimum time path. In practice we bound the maximum number of subsequent turn steps of (B.12) to $N_{\mathcal{T}} \leq \lceil \frac{2\pi}{\alpha} \rceil$, and we consider the shortest among the resulting feasible paths. Once $\tilde{\mathbf{u}}^*$ and the associated path $\tilde{X}^*$ have been determined, we transform back to the constant wind fixed target scenario by considering the control input sequence $\mathbf{u}^*$:

$$u^*(k) = \tilde{u}^*(k) + T_{u^*(k)} \cdot V_W \quad \forall (k = 1, \ldots, N-1), \tag{B.18}$$

where, by denoting with $v_a$ the MAV air speed,

$$T_{u^*(k)} = \frac{\|\tilde{u}^*(k)\|}{v_a} \tag{B.19}$$

is the time to execute step $u^*(k)$, assuming perfect path following. Using (B.14)–(B.17), it is trivial to show that $\mathbf{u}^*$ is feasible w.r.t. (B.5). Moreover, by (B.19), the tour time $T = \sum_{k=1}^{N-1} T_{u^*(k)}$, to execute the path $X^*$ generated by $\mathbf{u}^*$, is exactly the same time required to accomplish the associated path $\tilde{X}^*$ in the scenario of a moving target and null wind; this property holds in general for any path generated in the moving target scenario and transformed to the wind scenario by (B.18). Hence it follows that $X^*$ is of minimum time among the paths generated according to (B.9), in the scenario of zero wind and moving target, and transformed by (B.18).

In Fig. B.11 the effect of transformation (B.19) is depicted. Note that while turning against the wind results in a shorter step and a more gentle turn w.r.t. the original input $\tilde{u}^*(k)$, turning with the wind results in a longer step and a sharper turn. This effect accommodates the transient response of the MAV in wind (heavy solid line). This behavior is less damped when turning with the direction of the wind, and when turning against wind, it exhibits a highly damped response, allowing the next step to take effect sooner.

An example of the path planning method is depicted in Fig. B.12. The dashed path is the minimum time path generated with input of class (B.9), computed for the system in a zero wind field and a moving target. The solid line is the minimum time path transformed back to the system of constant wind and fixed target. The dotted regular octagon segments represent the maximum rate right turns $\mathcal{T}_{\text{init}}$ and $\mathcal{T}_{\text{final}}$.

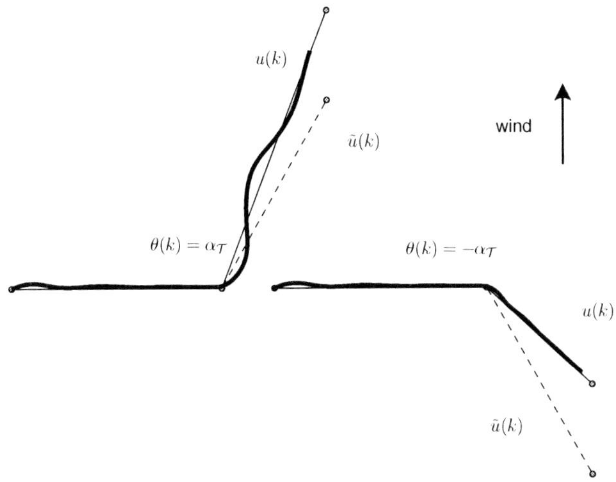


**Figure B.11:** Transformation of inputs from no-wind moving-target scenario ($\tilde{u}(k)$) to wind fixed-target scenario ($u(k)$) and associated transient response.

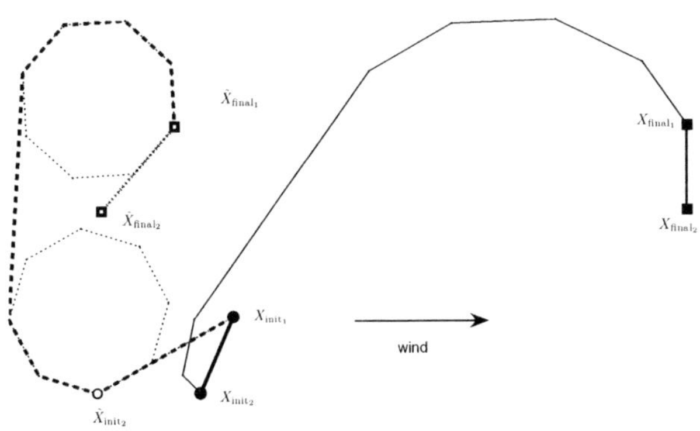


**Figure B.12:** Path planner between two pairs of waypoints.

## B.6 Summary

UAVs fly under the control of guidance navigation and control units. The user is responsible for generating paths that are consistent with the mission requirements, feasible with respect to the vehicles dynamics constraints, and suitable for the specific path following module running in the autopilot. A brief review on common autopilot path following approaches has been presented in section B.2. Vehicles models suitable for path designing have been discussed in section B.3. In section B.4 the Dubins [4] algorithm for minimum length path has been discussed along with guidelines for possible implementations. Finally an explicit waypoints path planning, in presence of wind, has been presented in section B.5.

# Bibliography

[1] R. Beard, D. Kingston, M. Quigley, D. Snyder, R. Christiansen, W. Johnson, T. McLain, and M. Goodrich. Autonomous vehicle technologies for small fixed-wing UAVs. *Journal of Aerospace Computing, Information, and Communication*, 2(1):92–108, 2005.

[2] X.-N. Bui, J.-D. Boissonnat, P. Soueres, and J.-P. Laumond. Shortest path synthesis for Dubins non-holonomic robot. In 1994 *IEEE International Conference on Robotics and Automation*, vol. 1, May 1994, 2–7.

[3] N. Ceccarelli, J. Enright, E. Frazzoli, S. Rasmussen, and C. Schumacher. Micro UAV path planning for reconnaissance in wind. In *Proceedings of the* 2007 *American Control Conference*, New York, July 2007.

[4] L. E. Dubins. On curves of minimal length with a constraint on average curvature, and with prescribed initial and terminal position. *American Journal of Mathematics*, 79(3):497–516, 1957.

[5] D. Gross, S. Rasmussen, P. Chandler, and G. Feitshans. Cooperative Operations in Urban TERrain (COUNTER). In *Proceedings of the SPIE Defense & Security Symposium*, Orlando, FL, April 2006.

[6] J.-C. Latombe. *Robot Motion Planning*. Springer, New York, 1991.

[7] C.-F. Lin. *Modern Navigation, Guidance, and Control Processing. Advanced Navigation, Guidance, and Control, and Their Applications*, Vol. II, Prentice–Hall, Englewood Cliffs, NJ, 1991.

[8] T. McGee, S. Spry, and J. Hedrick. Optimal path planning in a constant wind with a bounded turning rate. In *Proceedings of the* 2006 *AIAA Conference on Guidance, Navigation, and Control*, Keystone, CO, 2006.

[9] R. Murray and S. Sastry. Nonholonomic motion planning: Steering using sinusoids. *IEEE Transactions on Automatic Control*, 38(5):700–716, 1993.

[10] *Kestrel Autopilot System*. Procerus Technologies, www.procerusuav.com.

[11] B. Vaglienti and R. Hoag. *A Highly Integrated UAV Avionics System*. Cloud Cap Technology, www.cloudcaptech.com, 2003.

# Index